Springer Asia Pacific Mathematics Series

Volume 14

The Springer Asia Pacific Mathematics Series promotes high-quality scientific research connected to Asia-Pacific countries. Monographs, edited volumes, lecture notes, and textbooks will cover topics related to pure and applied mathematics as well as interdisciplinary research.

All material will be rigorously peer-reviewed and should be related to mathematical research conducted at or with institutions in the APAC region. This series is open to subseries, which may be topical or exclusively supplied by a single research institution.

Keonhee Lee · Carlos Morales · Ngocthach Nguyen

Topological Dynamics from a Measure-Theoretic Viewpoint

Springer

Keonhee Lee
Department of Mathematics
Chungnam National University
Daejeon, Korea (Republic of)

Carlos Morales
Hangzhou International Innovation Institute
of Beihang University
Hangzhou, China

Ngocthach Nguyen
Faculty of Mathematics
The University of Danang
Danang, Vietnam

ISSN 3091-2555 ISSN 3091-2563 (electronic)
Springer Asia Pacific Mathematics Series
ISBN 978-981-95-6389-0 ISBN 978-981-95-6390-6 (eBook)
https://doi.org/10.1007/978-981-95-6390-6

Mathematics Subject Classification: 37B25, 37B65, 37A05

This Springer imprint is published by the registered company Springer Nature Singapore Pte Ltd.
The registered company address is: 152 Beach Road, #21-01/04 Gateway East, Singapore 189721, Singapore

Preface

Classical measurable dynamics concerns itself with the asymptotic behavior of orbits within dynamical systems on measure spaces. A particular focus is placed on measure-preserving or ergodic dynamical systems. Central to this area are the Birkhoff ergodic and Poincaré recurrence theorems. Additional significant topics encompass the Krylov–Bogoliubov theorem, which addresses the existence of invariant and ergodic invariant measures, decay of correlations, mixing, Bernoulli or K-systems, and more.

However, a novel direction is emerging that examines topological dynamics through a measurable lens. To illustrate this, consider the example of a "minimal" topological dynamical system in which the only closed invariant sets are the empty set or the entire space. The natural question arises: what is the corresponding measurable concept? Analogous to the empty set and the entire space in topology, zero or full measure sets fulfill this role. Thus, we propose the definition of a "minimal measure" for a topological dynamical system, where every measurable invariant set has either zero or full measure. Interestingly, this aligns with the notion of an "ergodic measure." Therefore, we can view ergodicity as the measurable counterpart to minimality in topological dynamics.

Similarly, other concepts within topological dynamics can be explored from a measurable standpoint. For instance, a dynamical system on a metric space is considered "expansive" when the orbits of distinct points remain separated by a certain predefined radius. This definition can be relaxed by permitting a limited number of orbits to coexist with a given one. In measurable terms, this "fewness" corresponds to negligible terms. Specifically, we study Borel measures for which nearly all orbits separate within a specified radius. These measures are termed "expansive measures." Many properties of expansive dynamical systems, encompassing homeomorphisms and flows, extend to expansive measures. Chapters 3 and 7 delve into these aspects for homeomorphisms and flows on compact metric spaces, respectively.

This success has prompted some authors to translate other results from topological dynamics into measurable terms. An example is Walters' stability theorem, whose measurable version builds upon the notions of shadowing, expansive behavior, and topological stability measures. These adaptations are detailed in Chaps. 4 and 8 for homeomorphisms and flows, respectively. Another instance is Smale's spectral decomposition theorem, the measurable rendition of which is discussed in Chaps. 5 and 9 for homeomorphisms and flows.

In essence, our goal is to systematically present this measurable perspective on topological dynamics. We aspire to introduce this fresh outlook to researchers worldwide who are investigating topological dynamics from a measurable vantage point.

A part of this book was completed when the authors were visiting the Vietnam Institute for Advanced Study in Mathematics (VIASM) from October 2023 to January 2024. We thank the VIASM for their hospitality.

Daejeon, Korea (Republic of) Keonhee Lee
Hangzhou, China Carlos Morales
Danang, Vietnam Ngocthach Nguyen
2026

Contents

Part I
Discrete Dynamical Systems

Chapter 1
Preliminaries

In this chapter, we introduce the basic knowledge on measure theory for the development of this book. We first introduce the notions of Borel measures and integral of continuous functions on metric spaces. Moreover, we discuss the weak* topology on the collection of all Borel probability measures on a compact metric space. At the end we give a proof of the Krylov–Bogoliubov theorem.

1.1 Borel Measures and Integrals

Let X be a nonempty set and $\mathcal{P}(X)$ denote the power set of X. A collection $\Sigma \subset \mathcal{P}(X)$ is called a *σ-algebra* over X if it satisfies the following properties:

(1) $X, \emptyset \in \Sigma$;
(2) if $A \in \Sigma$, then $X \setminus A \in \Sigma$;
(3) if $\{A_i\}_{i \in \mathbb{N}}$ is a countable collection in Σ, then $\bigcup_{i \in \mathbb{N}} A_i \in \Sigma$.

For any nonempty collection $E \subset \mathcal{P}(X)$, there exists the unique smallest σ-algebra which contains the elements of E, which is called the *σ-algebra generated* by E.

Let Σ be a σ-algebra over X. A function $\mu : \Sigma \to \mathbb{R} \cup \{\infty\}$ is called a *measure* if it satisfies the following properties:

(1) $\mu(A) \geq 0$ for all $A \in \Sigma$;
(2) $\mu(\emptyset) = 0$;
(3) for a countable collection $\{A_i\}_{i \in \mathbb{N}}$ of pairwise disjoint sets in Σ,

$$\mu(\bigcup_{i \in \mathbb{N}} A_i) = \sum_{i \in \mathbb{N}} \mu(A_i).$$

If $\mu(X) = 1$, then μ is called a *probability measure* on X.

K. Lee et al., *Topological Dynamics from a Measure-Theoretic Viewpoint*, Springer Asia Pacific Mathematics Series 14, https://doi.org/10.1007/978-981-95-6390-6_1

Let (X, d) be a metric space. The *Borel σ-algebra* of X, denoted by β_X, is the σ-algebra generated by the collection of open sets of X. A *Borel measure* on X is a measure defined on the Borel σ-algebra β_X. The set of Borel probability measures on X is denoted by $\mathcal{M}(X)$. For a given $\mu \in \mathcal{M}(X)$, the support of μ, denoted by $\text{supp}(\mu)$, is defined as the collection of all points $x \in X$ satisfying that $\mu(U) > 0$ for every open sets U containing x.

The following proposition is a classical result which states the regularity of Borel measures on a metric space. We omit the proof.

Proposition 1.1 *Let μ be a Borel measure on a metric space X. Then for every Borel set $A \subset X$ and every $\varepsilon > 0$, there exist an open set U and a closed set C such that $C \subset A \subset U$ and $\mu(U \setminus C) < \varepsilon$.*

A function $f : X \to \mathbb{R}$ is called *measurable* if $f^{-1}(O) \in \beta_X$ for every open set $O \subset \mathbb{R}$. We observe that every continuous function $f : X \to \mathbb{R}$ is measurable.

We now introduce the concept of the integral of a continuous function on a compact metric space X with a Borel measure μ. For this, we recall some notions. A function $\psi : X \to \mathbb{R}$ is called a *simple* function if there exist a finite sequence of disjoint Borel sets $\{A_1, \cdots, A_n\}$ in β_X and a finite sequence of real numbers $\{a_1, \cdots, a_n\}$ such that

$$\psi = \sum_{i=1}^{n} a_i \chi_{A_i},$$

where $\chi_A : X \to \mathbb{R}$ is the characteristic function given by

$$\chi_A(x) = \begin{cases} 1 & \text{if } x \in A, \\ 0 & \text{otherwise.} \end{cases}$$

We define the integral of a simple function by

$$\int_X \psi d\mu = \sum_{i=1}^{n} a_i \mu(A_i).$$

Let $f : X \to \mathbb{R}$ be a continuous nonnegative function. Then the integral of f with respect to μ is defined by

$$\int_X f d\mu = \sup\{\int_X \psi d\mu : \psi \text{ is a simple function s.t } \psi \leq f\}.$$

Let $f : X \to \mathbb{R}$ be continuous function. Define the functions $f^+, f^- : X \to \mathbb{R}$ by

$$f^+(x) = \max\{f(x), 0\} \text{ and } f^-(x) = \max\{-f(x), 0\}.$$

Then the integral of f is defined by

$$\int_X f d\mu = \int_X f^+ d\mu - \int_X f^- d\mu.$$

In the rest of this section, we give a characterization of a Borel probability measure on a compact metric space. Let X be a compact metric space, and denote by $C(X)$ the space of all real-valued continuous functions on X. For any $f \in C(X)$, define the norm $\|f\| = \sup_{x \in X} |f(x)|$. We note that $C(X)$, equipped with the norm $\|\cdot\|$, is a separable Banach space. A linear operator $\varphi : C(X) \to \mathbb{R}$ is said to be *nonnegative* if $\varphi(f) \geq 0$ for every nonnegative function $f \in C(X)$. We say that φ is *normalized* if $\varphi(\mathbf{1}) = 1$, where $\mathbf{1}$ denotes the constant map on X given by $\mathbf{1}(x) = 1$ for every $x \in X$.

For any $\mu \in \mathcal{M}(X)$, define the map $\varphi_\mu : C(X) \to \mathbb{R}$ by

$$\varphi_\mu(f) = \int_X f d\mu, \quad \forall f \in C(X).$$

It is easy to verify that φ_μ is a normalized, nonnegative linear operator. Remarkably, the converse is also true: for any normalized, nonnegative linear operator $\varphi : C(X) \to \mathbb{R}$, there exists a unique measure $\mu \in \mathcal{M}(X)$ such that $\varphi = \varphi_\mu$. This result is known as the Riesz representation theorem, which plays an important role in measure theory and functional analysis. We state the theorem more precisely.

Theorem 1.1 *For any normalized nonnegative linear operator $\varphi : C(X) \to \mathbb{R}$, there is unique $\mu \in \mathcal{M}(X)$ such that*

$$\varphi(f) = \int_X f d\mu, \quad \forall f \in C(X). \tag{1.1}$$

The proof of this theorem is beyond the scope of this book. We refer the reader to [32] for details. We note that in Theorem 1.1, the measure $\mu \in \mathcal{M}(X)$ satisfying (1.1) is unique, which can be deduced by the following lemma.

Lemma 1.1 *Let μ and ν be two measures in $\mathcal{M}(X)$. If*

$$\int_X f d\mu = \int_X f d\nu, \quad \forall f \in C(X),$$

then $\mu = \nu$.

Proof. We first show that $\mu(C) = \nu(C)$ for all closed set C. In fact, let C be a closed set in X. For each $n \in \mathbb{N}$, we write

$$U_n = \{x \in X : d(x, C) < \frac{1}{n}\}.$$

It is clear that each U_n is open in X, and $C = \bigcap_{n \in \mathbb{N}} U_n$. Since C and U_n^c (the complement of U_n) are disjoint closed sets, by Urysohn's lemma, there exists a continuous function $f_n : X \to [0, 1]$ such that $f_n(x) = 0$ for $x \in U_n^c$ and $f_n(x) = 1$ for $x \in C$. Then we have

$$\mu(C) \le \int_X f_n d\mu = \int_X f_n d\nu = \int_{U_n} f_n d\nu \le \nu(U_n).$$

It implies that $\mu(C) \le \nu(C)$. Similarly, we also get $\nu(C) \le \mu(C)$. Therefore, $\mu(C) = \nu(C)$ for all closed set C in X. Consequently, we also have $\mu(U) = \nu(U)$ for every open set U.

Next, let $A \in \beta_X$ be arbitrary. By Proposition 1.1, for any $\varepsilon > 0$, there exist a closed set C_ε and an open set U_ε such that $C_\varepsilon \subset A \subset U_\varepsilon$, and

$$\mu(U_\varepsilon \setminus C_\varepsilon) < \varepsilon, \quad \nu(U_\varepsilon \setminus C_\varepsilon) < \varepsilon.$$

Since $\mu(C_\varepsilon) = \nu(C_\varepsilon)$ and $\mu(U_\varepsilon) = \nu(U_\varepsilon)$, it follows that

$$|\mu(A) - \nu(A)| \le \max\{\mu(U_\varepsilon) - \mu(C_\varepsilon), \nu(U_\varepsilon) - \nu(C_\varepsilon)\} < \varepsilon.$$

Since $\varepsilon > 0$ is arbitrary, we obtain that $\mu(A) = \nu(A)$. This completes the proof. □

1.2 The Weak* Topology

Let $\xi = \{f_i\}_{i\in\mathbb{N}}$ be a countable dense subset of $C(X)$. We define a map $d_\xi : \mathcal{M}(X) \times \mathcal{M}(X) \to \mathbb{R}$ by

$$d_\xi(\mu, \nu) = \sum_{i=1}^{\infty} \frac{1}{2^i \, \|f_i\|} \left| \int_X f_i \, d\mu - \int_X f_i \, d\nu \right|, \qquad \forall \mu, \nu \in \mathcal{M}(X). \tag{1.2}$$

Since

$$\frac{1}{\|f_i\|} \left| \int_X f_i \, d\mu - \int_X f_i \, d\nu \right| \le 2, \qquad \forall i \in \mathbb{N},$$

the series in (1.2) converges, and d_ξ is a well-defined metric on $\mathcal{M}(X)$. Moreover, if $\overline{\xi} = \{g_i\}_{i\in\mathbb{N}}$ is another countable dense subset of $C(X)$, then the corresponding metric $d_{\overline{\xi}}$ is equivalent to d_ξ, i.e., $d_{\overline{\xi}}$ and d_ξ induce the same topology on $\mathcal{M}(X)$.

Definition 1.1 The topology on $\mathcal{M}(X)$ induced by the metric d_ξ defined in (1.2) is called the *weak* topology* on $\mathcal{M}(X)$.

If we use the Riesz representation theorem, then we have the following theorem.

Theorem 1.2 *Suppose $\mathcal{M}(X)$ is equipped with the weak* topology. Then a sequence $\{\mu_n\}_{n\in\mathbb{N}}$ in $\mathcal{M}(X)$ converges to $\mu \in \mathcal{M}(X)$ if and only if*

$$\int_X f d\mu_n \to \int_X f d\mu \textit{ for every } f \in C(X).$$

Proof. First, we prove the direct implication. Let $\xi = \{f_n\}_{n\in\mathbb{N}}$ be a countable dense subset of the separable Banach space $C(X)$. Suppose that a sequence $\{\mu_n\}_{n\in\mathbb{N}}$ in $\mathcal{M}(X)$ converges to $\mu \in \mathcal{M}(X)$ in the weak* topology. Then we have

$$d_\xi(\mu_n, \mu) = \sum_{i=1}^{\infty} \frac{1}{2^i \|f_i\|} \left| \int_X f_i d\mu_n - \int_X f_i d\mu \right| \to 0 \text{ as } n \to \infty.$$

Hence, for each $i \in \mathbb{N}$, we get

$$\left| \int_X f_i d\mu_n - \int_X f_i d\mu \right| \to 0 \text{ as } n \to \infty.$$

Let $f \in C(X)$, and fix $\varepsilon > 0$. Since ξ is dense in $C(X)$, there exists $f_i \in \xi$ such that $\|f_i - f\| < \frac{\varepsilon}{4}$. Choose $N \in N$ such that $|\int_X f_i d\mu_n - \int_X f_i d\mu| < \frac{\varepsilon}{2}$ for all $n \geq N$. Then for $n \geq N$, we have

$$\begin{aligned}\left| \int_X f\, d\mu_n - \int_X f\, d\mu \right| &\leq \left| \int_X f\, d\mu_n - \int_X f_i\, d\mu_n \right| + \left| \int_X f_i\, d\mu_n - \int_X f_i\, d\mu \right| \\ &\quad + \left| \int_X f_i\, d\mu - \int_X f\, d\mu \right| \\ &\leq 2\|f_i - f\| + \frac{\varepsilon}{2} < \varepsilon.\end{aligned}$$

Since $\varepsilon > 0$ is arbitrary, it follows that

$$\left| \int_X f d\mu_n - \int_X f d\mu \right| \to 0 \text{ as } n \to \infty.$$

Conversely, suppose that for every $f \in C(X)$,

$$\int_X f d\mu_n \to \int_X f d\mu \text{ as } n \to \infty.$$

Fix $\varepsilon > 0$. Take $k \in \mathbb{N}$ such that

$$\sum_{i=k}^{\infty} \frac{1}{2^i \|f_i\|} \left| \int_X f_i d\mu_n - \int_X f_i d\mu \right| < \frac{\varepsilon}{2}$$

for all $n \in \mathbb{N}$. For this k, take $N \in \mathbb{N}$ satisfying

$$\sum_{i=1}^{k} \frac{1}{2^i \|f_i\|} \left| \int_X f_i d\mu_n - \int_X f_i d\mu \right| < \frac{\varepsilon}{2},\ \forall n \geq N.$$

Then, for any $n \geq N$ we have

$$
\begin{aligned}
d_\xi(\mu_n, \mu) &\leq \sum_{i=1}^{k} \frac{1}{2^i \, \|f_i\|} \left| \int_X f_i \, d\mu_n - \int_X f_i \, d\mu \right| \\
&\quad + \sum_{i=k}^{\infty} \frac{1}{2^i \, \|f_i\|} \left| \int_X f_i \, d\mu_n - \int_X f_i \, d\mu \right| \\
&< \varepsilon.
\end{aligned}
$$

This shows that μ_n converges to μ in the metric d_ξ and hence in the weak* topology. □

Remark 1.1 We say that a sequence $\{\mu_n\}_{n\in\mathbb{N}}$ in $\mathcal{M}(X)$ converges weakly to $\mu \in \mathcal{M}(X)$ (denoted as $\mu_n \to \mu$ weakly) if

$$\int_X f d\mu_n \to \int_X f d\mu \text{ for every } f \in C(X).$$

Hence the weak* topology on $\mathcal{M}(X)$ is the topology generated by the weak convergence.

Moreover, note that weak convergence $\mu_n \to \mu$ in $\mathcal{M}(X)$ does not imply that $\mu_n(A) \to \mu(A)$ for all $A \in \beta_X$. In fact, let $\{x_n\}_{n\in\mathbb{N}}$ be a sequence in X such that $x_n \to x \in X$ and $x_n \neq x$ for all $n \in \mathbb{N}$. Define the Borel measure $\delta_z : \beta_X \to \mathbb{R}$ by

$$\delta_z(A) = \begin{cases} 0 & \text{if } z \notin A, \\ 1 & \text{if } z \in A. \end{cases}$$

for all $A \in \beta_X$. The measure δ_z is called the *Dirac measure* centered at $z \in X$. Let $\mu_n = \delta_{x_n}$ and $\mu = \delta_x$. Then we have $\mu_n \to \mu$ weakly in $\mathcal{M}(X)$.

Let $A = \{x\} \in \beta_X$. Then for each $n \in \mathbb{N}$, we see that $\mu_n(A) = 0$ since $x_n \notin A$, and $\mu(A) = 1$ since $x \in B$. Hence $\mu_n(A)$ does not converge to $\mu(A)$.

Theorem 1.3 *Let $\{\mu_n\}_{n\in\mathbb{N}}$ be a sequence in $\mathcal{M}(X)$, and let $\mu \in \mathcal{M}(X)$. Then the following statements are equivalent:*

(1) μ_n converges to μ in the weak topology.*
(2) For any $f \in C(X)$, $\lim_{n\to\infty} \int_X f d\mu_n = \int_X f d\mu$.
(3) For any closed set C, $\limsup_{n\to\infty} \mu_n(C) \leq \mu(C)$.
(4) For any open set U, $\liminf_{n\to\infty} \mu_n(U) \geq \mu(U)$.
(5) For any Borel set A with $\mu(\partial A) = 0$, $\lim_{n\to\infty} \mu_n(A) = \mu(A)$.

Proof. (1) $\Leftrightarrow$ (2). The proof follows directly from Theorem 1.2.

(2) $\Rightarrow$ (3). Let C be a closed set in X. For each $k \in \mathbb{N}$, let $U_k = \{x \in X : d(x, C) < \frac{1}{k}\}$. We note that $U_{k+1} \subset U_k$ for all $k \in \mathbb{N}$ and $C = \bigcap_{k\in\mathbb{N}} U_k$. Since C and $X \setminus U_k$ are disjoint closed sets, by Urysohn's lemma, there exists a continuous function $f_k : X \to [0, 1]$ such that

$$f_k(x) = \begin{cases} 1 & \text{if } x \in C, \\ 0 & \text{if } x \in X \setminus U_k. \end{cases}$$

For each $k \in \mathbb{N}$, we have

$$\limsup_{n\to\infty} \mu_n(C) \le \limsup_{n\to\infty} \int_X f_k d\mu_n = \int_X f_k d\mu < \mu(U_k).$$

By letting $k \to \infty$, we get $\limsup_{n\to\infty} \mu_n(C) \le \mu(C)$.

(3) $\Leftrightarrow$ (4). Let U be an open set in X, and let $C = X \setminus U$. Since $\mu(U) = 1 - \mu(C)$, we have

$$\liminf_{n\to\infty} \mu_n(U) \ge 1 - \limsup_{n\to\infty} \mu_n(C) \ge 1 - \mu(C) = \mu(U).$$

Similarly, we can show that (4) $\Rightarrow$ (3).

(3) and (4) $\Rightarrow$ (5). Let A be a Borel set in X with $\mu(\partial A) = 0$. Then we see that $\mu(A) = \mu(\overline{A}) = \mu(A^o)$, where $\overline{A}$ denotes the closure of A and A^o denotes the interior of A. Then we have

$$\limsup_{n\to\infty} \mu_n(A) \le \limsup_{n\to\infty} \mu_n(\overline{A}) \le \mu(\overline{A}) = \mu(A),$$

and

$$\liminf_{n\to\infty} \mu_n(A) \ge \liminf_{n\to\infty} \mu_n(A^o) \ge \mu(A^o) = \mu(A).$$

Hence $\lim_{n\to\infty} \mu_n(A) = \mu(A)$.

(5) $\Rightarrow$ (1). Let $f \in C(X)$. By Theorem 1.2, it is sufficient to show that

$$\lim_{n\to\infty} \int_X f d\mu_n = \int_X f d\mu.$$

Let $a, b \in \mathbb{R}$ be such that $f(X) \subset (a, b)$. Define

$$f_*\mu : \beta_{[a,b]} \to \mathbb{R} \text{ by } f_*\mu(A) = \mu(f^{-1}(A)) \text{ for all } A \in \beta_{[a,b]}.$$

We note that $f_*\mu$ is a Borel measure on $[a, b]$. A point $t \in [a, b]$ is called a mass of $f_*\mu$ if $f_*\mu(\{t\}) > 0$. We note that the number of masses of $f_*\mu$ is countable. Then for any $\varepsilon > 0$, we can take a finite sequence $a = t_1 < \cdots < t_{m+1} = b$ such that $t_{i+1} - t_i < \varepsilon$ and $f_*\mu(\{t_i\}) = 0$ for all $i \in \{1, \cdots, m\}$.

For each $i \in \{1, \cdots, m\}$, let $A_i = f^{-1}([t_i, t_{i+1}))$. We observe that $\{A_i\}_{i=1}^m$ are pairwise disjoint of Borel sets and $\bigcup_{i=1}^m A_i = X$. Since $\partial A_i \subset f^{-1}(\{t_i, t_{i+1}\})$, we see that $\mu(\partial A_i) = 0$ for all $i \in \{1, \cdots, m\}$. Then we get $\lim_{n\to\infty} \mu_n(A_i) = \mu(A_i)$.

Define a simple function $\psi : X \to \mathbb{R}$ by $\psi = \sum_{i=1}^m t_i \chi_{A_i}$. We observe that $0 \le f(x) - \psi(x) \le \varepsilon$ for all $x \in X$. Then we have

$$\begin{aligned}\left|\int_X f\,d\mu_n - \int_X f\,d\mu\right| &\le \int_X |f-\psi|\,d\mu_n + \left|\int_X \psi\,d\mu_n - \int_X \psi\,d\mu\right| \\ &\quad + \int_X |f-\psi|\,d\mu \\ &\le 2\varepsilon + \sum_{i=1}^m |\mu_n(A_i) - \mu(A_i)|\,|t_i|.\end{aligned}$$

By letting $n \to \infty$, we have

$$\limsup_{n\to\infty} \left|\int_X f\,d\mu_n - \int_X f\,d\mu\right| \le 2\varepsilon.$$

Since $\varepsilon > 0$ is arbitrary, we get

$$\lim_{n\to\infty} \int_X f\,d\mu_n = \int_X f\,d\mu,$$

which completes the proof.

□

Theorem 1.4 *Let X be a compact metric space. Then $\mathcal{M}(X)$ is compact in the weak* topology.*

Proof. For simplicity, for any $\mu \in \mathcal{M}(X)$ and $f \in C(X)$, we write $\mu(f) := \int_X f\,d\mu$. Let $\{\mu_n\}_{n\in\mathbb{N}}$ be a sequence in $\mathcal{M}(X)$. We claim that it has a convergent subsequence.

Let $\xi = \{f_i\}_{i\in\mathbb{N}}$ be a countable dense subset of $C(X)$. For $f_1 \in \xi$, we consider the sequence of real numbers $\{\mu_n(f_1)\}_{n\in\mathbb{N}}$. Since the sequence is bounded by $\|f_1\|$ for all $n \in \mathbb{N}$, there is a convergent subsequence $\{\mu_n^1(f_1)\}_{n\in\mathbb{N}}$ of $\{\mu_n(f_1)\}_{n\in\mathbb{N}}$. For $f_2 \in \xi$, since the sequence $\{\mu_n^1(f_2)\}_{n\in\mathbb{N}}$ is bounded by $\|f_2\|$, there is a convergent subsequence $\{\mu_n^2(f_2)\}_{n\in\mathbb{N}}$ of $\{\mu_n^1(f_2)\}_{n\in\mathbb{N}}$. Continuing this process, for each $i \in \mathbb{N}$, there is a convergent subsequence $\{\mu_n^i(f_i)\}_{n\in\mathbb{N}}$ of $\{\mu_n^{i-1}(f_i)\}_{n\in\mathbb{N}}$. We observe that the sequence $\{\mu_n^i(f_k)\}_{n\in\mathbb{N}}$ is convergent for all $k \in \{1, \cdots, i\}$. Now consider the diagonal subsequence $\{\mu_n^n\}_{n\in\mathbb{N}}$. We see that for each fixed $i \in \mathbb{N}$, the sequence $\{\mu_n^n(f_i)\}_{n\in\mathbb{N}}$ is convergent. Since ξ is dense in $C(X)$, the sequence $\{\mu_n^n(f)\}_{n\in\mathbb{N}}$ is convergent for every $f \in C(X)$.

We define a map $\varphi : C(X) \to \mathbb{R}$ by

$$\varphi(f) = \lim_{n\to\infty} \mu_n^n(f), \quad \forall f \in C(X).$$

It is clear that φ is a bounded linear operator. Moreover, since $\varphi(\mathbf{1}) = 1$, and $\varphi(f) \ge 0$ for every nonnegative function $f \in C(X)$, by Theorem 1.1, there exists a Borel probability measure $\mu \in \mathcal{M}(X)$ such that

$$\varphi(f) = \int_X f\,d\mu, \quad \forall f \in C(X).$$

Hence $\int_X f d\mu_n^n \to \int_X f d\mu$ for all $f \in C(X)$. By Theorem 1.2, we obtain that μ_n^n converges to μ in the weak* topology.

□

1.3 Atomic and Nonatomic Measures

In this section, we study the notions of atomic measures and nonatomic measures on a compact metric space. Let X be a compact metric space. We show that the set of all atomic measures is dense in the $\mathcal{M}(X)$ with the weak* topology. Moreover, we prove that if X is uncountable, then there exists a nonatomic Borel measure on X.

Definition 1.2 A point $x \in X$ is called an *atom* of $\mu \in \mathcal{M}(X)$ if $\mu(\{x\}) > 0$. We say that $\mu \in \mathcal{M}(X)$ is *atomic* if its support consists of atoms, and *nonatomic* if it has no atoms.

Let X be a compact metric space, and $x \in X$. The Dirac measure δ_x is a basic example for atomic measures with $\text{supp}(\delta_x) = \{x\}$.

Let $X = [0, 1]$. Then the Borel measure $\mu \in \mathcal{M}(X)$ defined by

$$\mu = \frac{1}{2}\delta_0 + \sum_{n=1}^{\infty} \frac{1}{2^{n+1}}\delta_{\frac{1}{n}}$$

is an atomic measure with $\text{supp}(\mu) = \{0\} \cup \{\frac{1}{n} : n \in \mathbb{N}\}$, infinitely many points.

The next lemma tells us that the set of measures with finite support is dense in $\mathcal{M}(X)$.

Theorem 1.5 *Let X be a compact metric space. Then, the set of all measures $\mu \in \mathcal{M}(X)$ with finite support is dense in $\mathcal{M}(X)$. In particular, the set of atomic measures on X is dense in $\mathcal{M}(X)$.*

Proof. Fix $\mu \in \mathcal{M}(X)$. For any $n \in \mathbb{N}$, there are $k_n \in \mathbb{N}$ distinct points $x_{n,1}, \cdots, x_{n,k_n}$, and pairwise disjoint Borel sets $A_{n,1}, \cdots, A_{n,k_n}$ of diameter less than $\frac{1}{n}$ such that $x_{n,i} \in A_{n,i}$ for all $i = 1, \cdots, k_n$, and

$$\bigcup_{i=1}^{k_n} A_{n,i} = X.$$

For each $n \in \mathbb{N}$, we define a Borel measure $\mu_n \in \mathcal{M}(X)$ by

$$\mu = \sum_{i=1}^{k_n} \mu(A_{n,i})\delta_{x_{n,i}}.$$

Let $g \in C(X)$. For any $n \in \mathbb{N}$ and $1 \leq i \leq k_n$, we define

$$\alpha_{n,i} = \inf\{g(x) : x \in A_{n,i}\} \quad \text{and} \quad \beta_{n,i} = \sup\{g(x) : x \in A_{n,i}\}.$$

Then

$$\begin{aligned}\left|\int_X g\,d\mu_n - \int_X g\,d\mu\right| &= \left|\sum_{i=1}^{k_n}\int_{A_{n,i}} (g(x) - g(x_{n,i}))\,d\mu\right| \\ &\leq \sup_{1\leq i\leq k_n} |\beta_{n,i} - \alpha_{n,i}| \longrightarrow 0, \quad n \to \infty.\end{aligned}$$

By the definition of weak* topology on $\mathcal{M}(X)$, we obtain that $\mu_n \to \mu$.

□

Denote by $\mathcal{M}^*(X)$ the collection of all nonatomic Borel probability measures on a compact metric space X. In the rest of this section, we prove that if X is uncountable, then $\mathcal{M}^*(X)$ is nonempty. We say that a point x in a metric space X is *isolated* if there is $\varepsilon > 0$ such that $B(x, \varepsilon) = \{y \in X : d(x, y) < \varepsilon\} = \{x\}$.

Lemma 1.2 *Let X be a compact metric space without isolated points. Then $\mathcal{M}^*(X)$ is dense in $\mathcal{M}(X)$. In particular, $\mathcal{M}^*(X) \neq \emptyset$.*

Proof. Given $\varepsilon > 0$, we define

$$C(\varepsilon) = \{\mu \in \mathcal{M}(X) : \exists x \in X \text{ such that } \mu(\{x\}) \geq \varepsilon\}.$$

We claim that $C(\varepsilon)$ closed. Indeed, let $\{\mu_n\}_{n\in\mathbb{N}}$ be a sequence in $C(\varepsilon)$ such that it is weakly convergent to μ for some $\mu \in \mathcal{M}(X)$. For each $n \in \mathbb{N}$, let $x_n \in X$ be such that $\mu_n(\{x_n\}) \geq \varepsilon$. Since X is compact, we may assume that $x_n \to x_0$ for some $x_0 \in X$ as $n \to \infty$. For any $k \in \mathbb{N}$, by Theorem 1.3, we get

$$\mu(\overline{B(x_0, k^{-1})}) \geq \limsup_{n\to\infty} \mu_n(\overline{B(x_0, k^{-1})}) \geq \limsup_{n\to\infty} \mu_n(\{x_n\}) \geq \varepsilon.$$

Then we have

$$\mu(\{x_0\}) = \lim_{k\to\infty} \mu(\overline{B(x_0, k^{-1})}) \geq \varepsilon.$$

This implies $\mu \in C(\varepsilon)$, and so $C(\varepsilon)$ is closed in $\mathcal{M}(X)$.

We claim that $C(\varepsilon)$ has empty interior. Otherwise, by Theorem 1.5 we can take a measure μ_0 in the interior of $C(\varepsilon)$ which has finite support. We write $\mathrm{supp}(\mu_0) = \{y_1, \cdots, y_k\}$ and $p_i = \mu_0(\{y_i\})$ for all $i = 1, \cdots, k$. Then,

$$\sum_{i=1}^{k} p_i = 1 \quad \text{and} \quad \mu_0 = \sum_{i=1}^{k} p_i\delta_{y_i}.$$

Let $n \in \mathbb{N}$ be large enough such that $\{B(y_i, \frac{1}{n}) : i = 1, \cdots, k\}$ are pairwise disjoint. Choose $r \in \mathbb{N}$ such that

$$\frac{1}{r} < \frac{\varepsilon}{2}.$$

Since X has no isolated points, for each $i \in \{1, \cdots, k\}$ we can take r distinct points

$$\{z^n_{i,j} : j = 1, \cdots, r\} \in B(y_i, \frac{1}{n}),$$

and define

$$\mu_n = \frac{1}{r}\sum_{i=1}^{k}\sum_{j=1}^{r} p_i \delta_{z^n_{i,j}}.$$

Since

$$\mu_n(X) = \frac{1}{r}\sum_{i=1}^{k} rp_i = \sum_{i=1}^{k} p_i = 1,$$

one has $\mu_n \in \mathcal{M}(X)$. Also, we get

$$\lim_{n\to\infty} z^n_{i,j} = y_i, \qquad \forall (i, j) \in \{1, \cdots, k\} \times \{1, \cdots, r\}).$$

Then, for all $g \in C(X)$, we have

$$\int_X g d\mu_n = \frac{1}{r}\sum_{i=1}^{k}\sum_{j=1}^{r} p_i g(z^n_{i,j}) \to \frac{1}{r}\sum_{i=1}^{k} rp_i g(y_i) = \sum_{i=1}^{k} p_i g(y_i) = \int_X g d\mu$$

so $\{\mu_n\}_{n\in\mathbb{N}}$ converges weakly to μ_0. Then $\mu_n \in C(\varepsilon)$ for all sufficiently large n, and so, $\mu(\{z\}) \geq \varepsilon$ for some $z \in X$. However, the definition of μ_n implies

$$\mu_n(\{z\}) \leq \frac{1}{r} < \frac{\varepsilon}{2} \qquad (\forall z \in X)$$

which is clearly absurd. Then, $C(\varepsilon)$ has an empty interior as claimed.

Combining both claims we have that the set $O_m = \mathcal{M}(X) \setminus C(\frac{1}{m})$ is open and dense in $\mathcal{M}(X)$ for all $m \in \mathbb{N}$. Henceforth $\mathcal{R} = \bigcap_{m\in\mathbb{N}} O_m$ is residual in $\mathcal{M}(X)$.

Now, $\mathcal{M}(X)$ is compact by Theorem 1.4, and then $\mathcal{R}$ is dense by the Baire category theorem. Since $\mu \in \mathcal{R}$ implies $\mu \notin C(\frac{1}{m})$ for all $m \in \mathbb{N}$, one has

$$\mu(\{z\}) \leq \frac{1}{m}, \qquad \forall z \in X, \forall m \in \mathbb{N}.$$

Then, $\mu(\{z\}) = 0$ for all $z \in X$, which shows that $\mathcal{R} \subset \mathcal{M}^*(X)$. This completes the proof.

□

Theorem 1.6 *Let X be an uncountable compact metric space. Then there exists a nonatomic measure on X.*

Proof. First, we find a nonempty compact subset Y of X having no isolated points. Define

$$Y = \{x \in X : B(x,r) \text{ is uncountable for all } r > 0\}.$$

We observe that Y is nonempty. Otherwise, for each $x \in X$ there is $r_x > 0$ such that $B(x, r_x)$ is countable. Since X is compact, there is a finite covering $\{B(x_1, r_{x_1}), \cdots, B(x_n, r_{x_n})\}$ for some $x_1, \cdots, x_n \in X$. This would imply that X is countable, a contradiction.

We also have that Y is closed. In fact, let $\{x_n\}_{n\in\mathbb{N}}$ be a sequence in Y such that $x_n \to x \in X$ as $n \to \infty$. For any $r > 0$, there is $n \in \mathbb{N}$ such that $B(x_n, \frac{r}{2}) \subset B(x,r)$. Since $x_n \in Y$, we get $B(x,r)$ is uncountable. Then $x \in Y$, and so Y is closed. Since X is compact and Y is closed, it follows that Y is compact.

Suppose that Y has an isolated point, that is, there are $y \in Y$ and a constant $r_0 > 0$ such that

$$B[y, r_0] \cap Y = \{y\},$$

where $B[y,r] = \{z \in X : d(y,z) \le r\}$ denotes the closed r-ball centered at y. Fix $N \in \mathbb{N}$ such that $\frac{1}{n} < r_0$ for every $n \ge N$. Suppose for a while that $X' = B[y, r_0] \setminus B(y, \frac{1}{n})$ is uncountable for some $n \ge N$. Since X' is compact, we can apply the previous argument to find $y' \in X'$ such that $B(y', r)$ is uncountable for all $r > 0$. Then, $y' \in Y$ so $y' \in B[y, r_0] \cap Y = \{y\}$ thus $y' = y$ contradicting $y' \notin B(y, \frac{1}{n})$. Therefore, $B[y, r_0] \setminus B(y, \frac{1}{n})$ is countable for all $n \ge N$. Since

$$B[y, r_0] = \{y\} \cup \bigcup_{n \ge N} \left(B[y, r_0] \setminus B\left(y, \tfrac{1}{n}\right)\right),$$

it follows that $B[y, r_0]$ is countable. This is impossible, since $y \in Y$ and by definition every ball centered at y is uncountable. We conclude that Y has no isolated points.

Then, by Lemma 1.2, there is $\nu \in \mathcal{M}^*(Y)$. Defining

$$\mu(B) = \nu(B \cap Y), \text{ for all Borel set } B \subset X,$$

we get $\mu \in \mathcal{M}^*(X)$, completing the proof. □

1.4 Invariant Measures

In this section, we show the existence of invariant measures for continuous maps on compact metric spaces, which is known as the Krylov–Bogoliubov theorem. More precisely, let f be a continuous map on a compact metric space X. We first give the definition of invariant measure for f as follows.

Definition 1.3 A measure $\mu \in \mathcal{M}(X)$ is said to be *invariant* for f if $\mu(A) = \mu(f^{-1}(A))$ for all $A \in \beta_X$.

We now show that for every continuous map f on a compact metric space X, there exists an invariant measure for f on X. For this, we need some notations and a lemma. For given $\mu \in \mathcal{M}(X)$, we define a map $f_* : \mathcal{M}(X) \to \mathcal{M}(X)$ by $f_*(\mu)(A) = \mu(f^{-1}(A))$ for all $A \in \beta_X$.

Lemma 1.3 *Let f be a continuous map on a compact metric space X, and $\mu \in \mathcal{M}(X)$. Then*

$$\int_X \varphi df_*(\mu) = \int_X (\varphi \circ f) d\mu, \qquad \forall \varphi \in C(X).$$

Proof. Let $\mu \in \mathcal{M}(X)$. We define $\Phi : C(X) \to \mathbb{R}$ by

$$\Phi(\varphi) = \int_X (\varphi \circ f) d\mu, \qquad \forall \varphi \in C(X).$$

Clearly Φ is linear, nonnegative and $\Phi(\mathbf{1}) = 1$. Then, by Theorem 1.1, there is $\nu \in \mathcal{M}(X)$ such that $\Phi(\varphi) = \int_X \varphi d\nu$ for all $\varphi \in C(X)$. It follows that

$$\int_X \varphi d\nu = \int_X (\varphi \circ f) d\mu, \qquad \forall \varphi \in C(X). \tag{1.3}$$

Let C be a closed subset of X, and a constant $\varepsilon > 0$. By Proposition 1.1, there exists an open set O in X such that $C \subset O$ and $\nu(O \setminus C) < \varepsilon$. By Urysohn's lemma there is $\varphi \in C(X)$ such that $0 \le \varphi \le 1$, $\varphi = 0$ on O^c and 1 in C. Then we have

$$\begin{aligned}
\nu(C) &= \int_X \chi_C \, d\nu = \int_X \varphi \, d\nu - \int_{O \setminus C} \varphi \, d\nu \\
&\overset{(1.3)}{\ge} \int_X (\varphi \circ f) \, d\mu - \nu(O \setminus C) \\
&\ge \int_X \chi_{f^{-1}(C)} \, d\mu - \varepsilon \\
&= \mu(f^{-1}(C)) - \varepsilon = f_*(\mu)(C) - \varepsilon.
\end{aligned}$$

Since ε is arbitrary, we get

$$\nu(C) \ge f_*(\mu)(C), \quad \forall C \subset X \text{ closed}.$$

Let $A \in \beta_X$ and $\varepsilon > 0$. By Proposition 1.1, there is $C \subset A$ closed such that $f_*(\mu)(A \setminus C) < \varepsilon$. Then we have

$$f_*(\mu)(A) = f_*(\mu)(C) + f_*(\mu)(A \setminus C) \le \nu(C) + \varepsilon \le \nu(A) + \varepsilon.$$

Since ε is arbitrary, we get

$$\nu(A) \geq f_*(\mu)(A), \quad \forall A \in \beta_X.$$

On the other hand, for any Borel set $A \in \beta_X$, we have

$$1 - \nu(A) = \nu(A^c) \geq f_*(\mu)(A^c) = 1 - f_*(\mu)(A).$$

It implies that

$$\nu(A) \leq f_*(\mu)(A), \quad \forall A \in \beta_X.$$

Hence $\nu(A) = f_*(\mu)(A)$ for all $A \in \beta_X$. By replacing $\nu = f_*(\mu)$ to (1.3), we complete the proof.

□

Denote by $\mathcal{M}(X, f)$ the collection of all invariant measures of f. Replacing $f_*(\mu) = \mu$ in Lemma 1.3, we get the following characterization.

Corollary 1.1 *Let X be a compact metric space, and let $f : X \to X$ be continuous. Then*

$$\mathcal{M}(X, f) = \left\{ \mu \in \mathcal{M}(X) : \int \varphi \, d\mu = \int (\varphi \circ f) \, d\mu \ \text{ for all } \varphi \in C(X) \right\}.$$

In the following theorem, we show that $\mathcal{M}(X, f)$ is nonempty compact subset of $\mathcal{M}(X)$.

Theorem 1.7 *Let f be a continuous map on a compact metric space X. Then $\mathcal{M}(X, f)$ is a nonempty compact subset of $\mathcal{M}(X)$.*

Proof. We first show that $\mathcal{M}(X, f)$ is nonempty. Let $\nu \in \mathcal{M}(X)$, and define a sequence $\{\nu_n\}_{n\in\mathbb{N}}$ in $\mathcal{M}(X)$ by

$$\nu_n = \frac{1}{n} \sum_{i=0}^{n-1} f_*^i(\nu).$$

Since $\mathcal{M}(X)$ is compact, there exist $\mu \in \mathcal{M}(X)$ and a subsequence ν_{n_k} of ν_n such that $\nu_{n_k} \to \mu$ in the weak* topology. We shall prove $\mu \in \mathcal{M}(X, f)$.

For any fixed $\varphi \in C(X)$, set

$$A_n(\varphi) := \frac{1}{n} \sum_{i=0}^{n-1} \int_X \varphi \circ f^i \, d\nu.$$

Then we have

$$\int_X \varphi \circ f \, d\nu_n = \frac{1}{n} \sum_{i=0}^{n-1} \int_X \varphi \circ f^{i+1} d\nu$$

$$= A_n(\varphi) + \frac{1}{n}\left(\int_X \varphi \circ f^n d\nu - \int_X \varphi\, d\nu\right).$$

Hence, for bounded φ,

$$\left|\int_X \varphi \circ f\, d\nu_n - A_n(\varphi)\right| \le \frac{2}{n}\, \|\varphi\| \to 0 \text{ as } n \to \infty.$$

If $\nu_{n_k} \to \mu$ in the weak* topology, then

$$\int_X \varphi \circ f\, d\mu = \lim_k \int_X \varphi \circ f\, d\nu_{n_k} = \lim_k A_{n_k}(\varphi)$$
$$= \lim_k \int_X \varphi\, d\nu_{n_k} = \int_X \varphi\, d\mu.$$

Therefore,

$$\int_X (\varphi \circ f) d\mu = \int_X \varphi d\mu, \quad \forall \varphi \in C(X),$$

and by Corollary 1.1, we conclude that $\mu \in \mathcal{M}(X, f)$. Thus, $\mathcal{M}(X, f)$ is nonempty.

Next, we show that $\mathcal{M}(X, f)$ is compact in $\mathcal{M}(X)$. Let $\{\mu_n\}_{n\in\mathbb{N}}$ be a sequence in $\mathcal{M}(X, f)$ such that $\{\mu_n\}_{n\in\mathbb{N}}$ is weakly convergent to μ for some $\mu \in \mathcal{M}(X)$. Then,

$$\lim_{n\to\infty} \int_X \varphi d\mu_n = \int_X \varphi d\mu, \quad \forall \varphi \in C(X).$$

By Corollary 1.1, we get

$$\int_X (\varphi \circ f) d\mu = \lim_{n\to\infty} \int_X (\varphi \circ f) d\mu_n$$
$$= \lim_{n\to\infty} \int_X \varphi d\mu_n = \int_X \varphi d\mu, \quad \forall \varphi \in C(X).$$

Hence, $\mu \in \mathcal{M}(X, f)$. Therefore, $\mathcal{M}(X, f)$ is compact.

□

Definition 1.4 A Borel measure μ of X is said to be *ergodic* for a continuous map on a compact metric space $f : X \to X$ if for any f-invariant Borel subset $A \in \beta_X$, $\mu(A) = 0$ or $\mu(A) = 1$.

We recall that A is *f-invariant* if $f^{-1}(A) = A$. Note that this definition does not need the invariance of the measure. We denote by $\mathcal{E}(X, f)$ the collection of all ergodic invariant measures of f. In the following theorem, we give a characterization of ergodic invariant measures, and show that $\mathcal{E}(X, f) \neq \emptyset$. For this, we need some notations. A measure μ is called an *extreme point* of $\mathcal{M}(X, f)$ if whenever $\mu = t\nu + (1-t)\gamma$ for some $0 \le t \le 1$ and $\nu, \gamma \in \mathcal{M}(X, f)$ then we have $\mu = \nu = \gamma$.

Given $\mu, \nu \in \mathcal{M}(X)$ we say that ν is absolutely continuous with respect to μ, denoted by $\nu << \mu$, if for every $A \in \beta_X$, $\mu(A) = 0$ implies $\nu(A) = 0$. Let

$$L^1_\mu(X) = \{\varphi : X \to \mathbb{R} : \int_X |\varphi| d\mu < \infty\}.$$

The Radon–Nikodym theorem asserts that if $\nu << \mu$, then there is $\varphi \in L^1_\mu(X)$ such that

$$\nu(A) = \int_A \varphi d\mu, \quad \forall A \in \beta_X.$$

This φ is called the *Radon–Nikodym derivative* of ν respect to μ, and denoted by $\frac{d\nu}{d\mu}$.

The following theorem provides a characterization of ergodic invariant measures as the extreme points of the space $\mathcal{M}(X, f)$ of invariant measures. This is consistent with the KreinâŁ"Milman theorem, which states that a compact convex set is determined by its extreme points.

Theorem 1.8 *Let f be a continuous map on a compact metric space X. Then μ is an extreme point of $\mathcal{M}(X, f)$ if and only if μ is an ergodic invariant measure of f. In particular, $\mathcal{E}(X, f) \neq \emptyset$.*

Proof. Let μ be an extreme point of $\mathcal{M}(X, f)$. Suppose that μ is not ergodic. Then there is an invariant set $E \in \beta_X$ such that $0 < \mu(E) < 1$. Define two invariant measures $\nu, \gamma \in \mathcal{M}(X, f)$ by

$$\nu(B) = \frac{\mu(E \cap B)}{\mu(E)} \quad \text{and} \quad \gamma(B) = \frac{\mu(E^c \cap B)}{\mu(E^c)}, \quad \forall B \in \beta_X.$$

Taking $t = \mu(E)$, we get $0 < t < 1$ and

$$\mu(B) = \mu(E)\frac{\mu(E \cap B) + \mu(E^c \cap B)}{\mu(E)} = t\nu(B) + (1-t)\gamma(B), \quad \forall B \in \beta_X.$$

Then $\mu = t\nu + (1-t)\gamma$. Since E is invariant, we get $\nu, \gamma \in \mathcal{M}(X, f)$, and so $\mu = \nu = \gamma$. In particular, $\mu = \nu$. Replacing $B = E$ in the definition of ν we get $\mu(E) = 1$ contradicting $0 < \mu(E) < 1$. Hence, μ is ergodic.

Conversely, suppose that μ is ergodic, and satisfies

$$\mu = t\nu + (1-t)\gamma \quad \text{for some} \quad \nu, \gamma \in \mathcal{M}(X, f) \quad \text{and} \quad 0 < t < 1. \tag{1.4}$$

It follows that $\nu << \mu$, and so there is the Radon–Nikodym derivative $\frac{d\nu}{d\mu}$ i.e.,

$$\nu(B) = \int_B \frac{d\nu}{d\mu} d\mu, \quad \forall B \in \beta_X.$$

We have $\frac{d\nu}{d\mu} \geq 0$ for μ-a.e. Define

$$E = \{x \in X : \frac{d\nu}{d\mu}(x) < 1\}.$$

Then we get

$$\int_{E \cap f^{-1}(E)} \frac{d\nu}{d\mu} d\mu + \int_{E \setminus f^{-1}(E)} \frac{d\nu}{d\mu} d\mu = \int_E \frac{d\nu}{d\mu} d\mu = \nu(E) = \nu(f^{-1}(E))$$

and

$$\nu(f^{-1}(E)) = \int_{f^{-1}(E) \cap E} \frac{d\nu}{d\mu} d\mu + \int_{f^{-1}(E) \setminus E} \frac{d\nu}{d\mu} d\mu,$$

and so

$$\int_{E \setminus f^{-1}(E)} \frac{d\nu}{d\mu} d\mu = \int_{f^{-1}(E) \setminus E} \frac{d\nu}{d\mu} d\mu.$$

Now, $\frac{d\nu}{d\mu} < 1$ on $E \setminus f^{-1}(E)$ and ≥ 1 on $f^{-1}(E) \setminus E$. Moreover,

$$\begin{aligned} \mu(f^{-1}(E) \setminus E) &= \mu(f^{-1}(E)) - \mu(f^{-1}(E) \cap E) \\ &= \mu(E) - \mu(f^{-1}(E) \cap E) \\ &= \mu(E \setminus f^{-1}(E)). \end{aligned}$$

Therefore, $\mu(f^{-1}(E) \setminus E) = \mu(E \setminus f^{-1}(E)) = 0$ proving

$$\mu(f^{-1}(E) \Delta E) = 0 \quad \text{(where } \Delta \text{ denotes symmetric difference of sets)}$$

and so $\mu(E) = 0$ or 1 since μ is ergodic. If $\mu(E) = 1$,

$$\nu(X) = \int_E \frac{d\nu}{d\mu} d\mu < \mu(E) = 1,$$

which is a contradiction. Hence $\mu(E) = 0$. Likewise, if $F = \{x \in X : \frac{d\nu}{d\mu}(x) > 1\}$, then $\mu(F) = 0$ proving $\frac{d\nu}{d\mu} = 1$ for μ-a.e. Then,

$$\nu(B) = \int_B \frac{d\nu}{d\mu} d\mu = \int_B d\mu = \mu(B), \quad \forall B \in \beta_X,$$

proving $\nu = \mu$. Replacing in (1.4) and using that $0 < t < 1$, we obtain $\mu = t\mu + (1 - t)\gamma$, which implies that $\mu = \gamma$.

To show that $\mathcal{E}(X, f)$ is nonempty, it is sufficient to prove that $\mathcal{M}(X, f)$ has an extreme point. Let $\{f_i\}_{i=0}^{\infty}$ be a countable dense subset of $C(X)$. For continuous map f_0, we consider $\varphi_0 : \mathcal{M}(X, f) \to \mathbb{R}$ given by

$$\varphi_0(\mu) = \int_X f_0 d\mu.$$

Since φ_0 is continuous and $\mathcal{M}(X, f)$ is compact, we see that

$$\mathcal{M}_0 = \{\nu \in \mathcal{M}(X, f) : \int_X f_0 d\nu = \sup_{\mu \in \mathcal{M}(X, f)} \int_X f_0 d\mu\}$$

is a nonempty compact subset of $\mathcal{M}(X, f)$.

Similarly, for continuous map f_1, we obtain that

$$\mathcal{M}_1 = \{\nu \in \mathcal{M}_0 : \int_X f_1 d\nu = \sup_{\mu \in \mathcal{M}_0} \int_X f_1 d\mu\}$$

is a nonempty compact subset of $\mathcal{M}_0$.

Continue the process, for each $i \geq 1$, we have

$$\mathcal{M}_i = \{\nu \in \mathcal{M}_{i-1} : \int_X f_i d\nu = \sup_{\mu \in \mathcal{M}_{i-1}} \int_X f_i d\mu\}$$

is a nonempty compact subset of $\mathcal{M}_{i-1}$. Hence we get a sequence $\{\mathcal{M}_i\}_{i \in \mathbb{N}}$ of nonempty compact sets such that

$$\mathcal{M}(X, f) \supset \mathcal{M}_0 \supset \mathcal{M}_1 \supset \cdots \mathcal{M}_i \supset \cdots$$

Let $\mathcal{M}_\infty = \bigcap_{i=0}^{\infty} \mathcal{M}_i$. We note that $\mathcal{M}_\infty$ is nonempty, and so we can take $\mu \in \mathcal{M}_\infty$. We will show that μ is an extreme point of $\mathcal{M}(X, f)$. Indeed, suppose that $\mu = t\mu_1 + (1-t)\mu_2$ for some $\mu_1, \mu_2 \in \mathcal{M}(X, f)$ and $0 \leq t \leq 1$. We claim that

$$\int_X f_i d\mu = \int_X f_i d\mu_1 = \int_X f_i d\mu_2, \ \forall i \geq 0.$$

For each $i \in \mathbb{N}$, since

$$\int_X f_i d\mu = t \int_X f_i d\mu_1 + (1-t) \int_X f_i d\mu_2,$$

we obtain

$$\int_X f_i d\mu \leq \max\left\{\int_X f_i d\mu_1, \int_X f_i d\mu_2\right\}. \tag{1.5}$$

For $i = 0$, since $\mu \in \mathcal{M}_0$, we have

$$\int_X f_0 d\mu \geq \max\{\int_X f_0 d\mu_1, \int_X f_0 d\mu_2\}.$$

Combining this with (1.5), it follows that

$$\int_X f_0 d\mu = \max\{\int_X f_0 d\mu_1, \int_X f_0 d\mu_2\}.$$

Since $\mu = t\mu_1 + (1-t)\mu_2$, we conclude that

$$\int_X f_0 d\mu = \int_X f_0 d\mu_1 = \int_X f_0 d\mu_2.$$

Assume that for some $i \geq 0$

$$\int_X f_i d\mu = \int_X f_i d\mu_1 = \int_X f_i d\mu_2.$$

It implies that $\mu_1, \mu_2 \in \mathcal{M}_i$. Since $\mu \in \mathcal{M}_{i+1}$, we have

$$\int_X f_{i+1} d\mu \geq \max\{\int_X f_{i+1} d\mu_1, \int_X f_{i+1} d\mu_2\}.$$

Then by (1.5), we obtain

$$\int_X f_{i+1} d\mu = \max\{\int_X f_{i+1} d\mu_1, \int_X f_{i+1} d\mu_2\}.$$

Using again $\mu = t\mu_1 + (1-t)\mu_2$, it follows that

$$\int_X f_{i+1} d\mu = \int_X f_{i+1} d\mu_1 = \int_X f_{i+1} d\mu_2.$$

By mathematical induction, we conclude that

$$\int_X f_i d\mu = \int_X f_i d\mu_1 = \int_X f_i d\mu_2, \ \forall i \geq 0.$$

Moreover, since $\{f_i\}_{i\in\mathbb{Z}}$ is dense in $C(X)$, we deduce that

$$\int_X f d\mu = \int_X f d\mu_1 = \int_X f d\mu_2, \ \forall f \in C(X).$$

By Lemma 1.1, we derive that $\mu = \mu_1 = \mu_2$, which means that μ is an extreme point of $\mathcal{M}(X, f)$. Then, $\mu \in \mathcal{E}(X, f)$ by the first part of the theorem. Therefore, $\mathcal{E}(X, f) \neq \emptyset$ completing the proof.

□

Chapter 2
Topological Dynamics

In this chapter, we study the basic knowledge in topological dynamics. Through these we will prove two important results: Walters' stability and Smale's spectral decomposition theorems.

2.1 Basic Concepts

Let f be a homeomorphism on a compact metric space X. For any $x \in X$, the set $\{f^n(x)\}_{n\in\mathbb{Z}}$ is called the *orbit* of x under f, we denote it by $\mathcal{O}_f(x)$.

Definition 2.1 A point $x \in X$ is said to be:

(1) *fixed* for f if $f(x) = x$. We denote by $Fix(f)$ the collection of all fixed points of f;
(2) *periodic* for f if $f^n(x) = x$ for some $n \geq 1$. The minimal positive integer n satisfies the equality is called the *period* of x. We denote by $Per(f)$ the collection of all periodic points of f; The orbit of $x \in Per(f)$ is called a *periodic orbit* of f.

A subset $A \subset X$ is said to be *invariant* for f (or f-invariant) if $f^{-1}(A) = A$, which is equivalent to $f(A) = A$ since f is a homeomorphism. Note that any orbit of f is invariant for f. Moreover, A is invariant if and only if it is a union of orbits.

Definition 2.2 A nonempty set $A \subset X$ is said to be *minimal* for f if it is compact invariant and contains no closed invariant sets other than $\emptyset$ and A.

Equivalently, A is minimal if it is compact invariant and has no proper compact invariant subset of f. Using Zorn's lemma, we can show that any nonempty compact invariant set contains a minimal set. Let us present it precisely as follows.

Theorem 2.1 *Let f be a homeomorphism on a compact metric space X. Then there exists a minimal set A of f.*

K. Lee et al., *Topological Dynamics from a Measure-Theoretic Viewpoint*, Springer Asia Pacific Mathematics Series 14, https://doi.org/10.1007/978-981-95-6390-6_2

Definition 2.3 A point $x \in X$ is said to be *minimal* for f if it belongs to a minimal set of f, and *nonminimal* otherwise.

Definition 2.4 A point $y \in X$ is said to be an *ω-limit point* (resp. *α-limit point*) of $x \in X$ if there exists a sequence $n_i \to +\infty$ (resp. $n_i \to -\infty$) such that $f^{n_i}(x) \to y$. Denote the set of all ω-limit points (resp. α-limit points) of x by $\omega(x, f)$ (resp. $\alpha(x, f)$).

Theorem 2.2 *Let f be a homeomorphism on a compact metric space X. For any $x \in X$, the sets $\omega(x, f)$ and $\alpha(x, f)$ are nonempty, compact and invariant.*

Theorem 2.3 *Let f be a homeomorphism on a compact metric space X, and Λ be a subset of X. Then the following properties are equivalent:*

(1) *Λ is minimal;*
(2) *$\omega(x, f) = \Lambda$ for every $x \in \Lambda$;*
(3) *$\mathcal{O}_f(x)$ is dense in Λ for every $x \in \Lambda$.*

Proof. (1) $\Rightarrow$ (2). Let Λ be a minimal set of f, and $x \in \Lambda$. Since $\omega(x, f)$ is a nonempty compact invariant subset of Λ, one has $\omega(x, f) = \Lambda$.

(2) $\Rightarrow$ (3). Let $x \in \Lambda$. Clearly $\omega(x, f) \subset \overline{\mathcal{O}_f(x)}$. Then $\mathcal{O}_f(x)$ is dense in Λ.

(3) $\Rightarrow$ (1). Suppose that Λ is not minimal. Then there is a compact invariant subset $\emptyset \neq \Lambda_1 \neq \Lambda$. Since $\Lambda_1 \neq \emptyset$ we can choose $x \in \Lambda_1$. We see that $\overline{\mathcal{O}_f(x)} \subset \Lambda_1$, and so $\overline{\mathcal{O}_f(x)} \neq \Lambda$. In particular, $\mathcal{O}_f(x)$ is not dense in Λ. The contradiction completes the proof.

□

Corollary 2.1 *Let f be a homeomorphism on a compact metric space X. Then $x \in X$ is minimal for f if and only if $\omega(x, f)$ contains x and is a minimal set of f.*

The main objects in topological dynamics are various types of recurrence of orbits. We see that periodic orbit has the strongest recurrence. Besides it, there are several weaker notions of recurrence of orbits.

Definition 2.5 A point $x \in X$ is said to be *recurrent* for f if $x \in \omega(x, f) \cup \alpha(x, f)$. The set of all recurrent points of f is called the recurrent set of f, denoted by $R(f)$.

By Corollary 2.1, every minimal point is recurrent for f. Then, by Theorem 2.1, it follows that every homeomorphism on a compact metric space admits a recurrent point.

Theorem 2.4 *Let f be a homeomorphism on a compact metric space X. The set $R(f)$ is nonempty and invariant, but it may be noncompact.*

Definition 2.6 A point $x \in X$ is said to be *nonwandering* for f if for any neighborhood U of x, there is $n \geq 1$ such that $f^n(U) \cap U \neq \emptyset$. The set of all nonwandering points of f is called the *nonwandering set* of f, denoted by $\Omega(f)$.

Definition 2.7 (1) For any $\delta > 0$, a sequence $\{x_i\}_{i=a}^{b}$ $(-\infty \le a < b \le \infty)$ in X is called a *δ-pseudo orbit* (or *δ-chain*) of f if $d(f(x_i), x_{i+1}) \le \delta$ for all $a \le i \le b-1$. We say that a finite sequence $\{x_i : 0 \le i \le n\}$ is a δ-chain of f from x to y if it is a δ-chain, $x_0 = x$ and $x_n = y$.

(2) A point $x \in X$ is a *chain recurrent point* of f if for any $\varepsilon > 0$ there exists ε-chain $\{x_i\}_{i=0}^{n}$ of f from x to itself. The collection of all chain recurrent points of f is called the *chain recurrent set* of f, denoted by $CR(f)$.

Remark 2.1 Let f be a homeomorphism on a compact metric space X. We observe that the chain recurrent set of the restriction $f|_{CR(f)}$ coincides with $CR(f)$. More precisely, a point $x \in CR(f)$ if and only if for any $\varepsilon > 0$ there exists an ε-chain $\{x_i\}_{i=0}^{n} \subset CR(f)$ from x to itself (see, for example, Theorem 3.1.6 [3]).

Theorem 2.5 *Let f be a homeomorphism on a compact metric space. Then:*

(1) $\overline{Per(f)} \subset \Omega(f) \subset CR(f)$*;*
(2) $\Omega(f)$ *and* $CR(f)$ *are compact and invariant, but* $Per(f)$ *is not closed in general;*
(3)

$$\bigcup_{x \in X} \omega(x, f) \subset \Omega(f).$$

We now recall two fundamental concepts in topological dynamics, which are expansiveness and shadowing for homeomorphisms on compact metric spaces.

Definition 2.8 We say that a homeomorphism f on a compact metric space X is *expansive* on an invariant subset $\Lambda \subset X$ if there is $\delta > 0$ such that if $x, y \in \Lambda$ satisfy $d(f^n(x), f^n(y)) \le \delta$ for every $n \in \mathbb{Z}$, then $x = y$. Here δ is called an *expansive constant* of f. A homeomorphism f is said to be expansive if f is expansive on X.

Definition 2.9 We say that a homeomorphism f on a compact metric space X is *shadowing* on an invariant subset $\Lambda \subset X$ if for every $\varepsilon > 0$ there is $\delta > 0$ such that every δ-pseudo orbit $\{x_i\}_{i \in \mathbb{Z}} \subset \Lambda$ of f is ε-shadowed by a point $x \in X$, that is,

$$d(f^i(x), x_i) \le \varepsilon, \ \forall i \in \mathbb{Z}.$$

We say that f is *shadowing* if it is shadowing on X.

Remark 2.2 In the following, we provide a characterization of shadowing homeomorphisms on compact metric spaces. In fact, it is easy to show that the following statements are equivalent:

(1) f is shadowing;
(2) for every $\varepsilon > 0$, there is $\delta > 0$ such that every finite δ-pseudo orbit $\{x_i\}_{i=a}^{b}$ $(-\infty < a < b < \infty)$ is ε-shadowed, i.e., there is $x \in X$ such that $d(f^i(x), x_i) \le \varepsilon$ for $a \le i \le b$.

We say that f is *finite shadowing* if it satisfies item (2).

For any $x, y \in X$ and $\delta > 0$, we say that $x \sim^\delta y$ if there are a δ-chain from x to y and a δ-chain from y to x. If we define $x \sim y$ whenever $x \sim^\delta y$ for all $\delta > 0$, then we see that $\sim$ gives an equivalence relation on $CR(f)$. An equivalence class of $\sim$ is called a *chain component* of f. It is clear that every chain component of f is closed and invariant.

Lemma 2.1 *Let f be a homeomophism on a compact metric space X. If $d(x, y) < \delta$ for $x, y \in CR(f)$ and $\delta > 0$, then $x \sim^\delta y$.*

Proof. Let $\delta > 0$, and $x, y \in CR(f)$ be such that $d(x, y) < \delta$. Take $\varepsilon = \delta - d(x, y)$. Since $x \in CR(\phi)$, there is an ε-chain $\{x_0, \cdots, x_n\}$ from x to itself. Since

$$d(f(x_{n-1}), y) \leq d(f(x_{n-1}), x) + d(x, y) < \delta,$$

we see that $\{x_0, \cdots, x_{n-1}, y\}$ is a δ-chain from x to y. Similarly, we can construct a δ-chain from y to x. Hence we obtain $x \sim^\delta y$.

□

Theorem 2.6 *Let f be a shadowing homeomorphism on a compact metric space X. Then $\Omega(f) = CR(f)$, and the restriction $f|_{\Omega(f)}$ is shadowing.*

Proof. Let $x \in CR(f)$ and $\varepsilon > 0$. By the shadowing of f, we take $\delta > 0$ such that every δ-pseudo orbit of f is ε-shadowed by a point in X. Since $x \in CR(f)$, there is a δ-chain $\xi = \{x_i\}_{i=1}^n$ from x to itself. Then there is $y \in X$ such that $d(y, x) < \varepsilon$ and $d(f^n(y), x) < \varepsilon$. It implies that $f^n(B(x, \varepsilon)) \cap B(x, \varepsilon) \neq \emptyset$. Since $\varepsilon > 0$ is arbitrary, we get $x \in CR(f)$, and so $CR(f) = \Omega(f)$.

We now prove that the restriction $f|_{\Omega(f)}$ is shadowing. Let $\varepsilon > 0$, and take $\delta > 0$ such that every δ-pseudo orbit of f is ε-shadowed by a point in X. We can split $\Omega(f) = CR(f) = \bigcup_\lambda A_\lambda$, where A_λ is an equivalent class with respect to the relation $\sim^\delta$. For an equivalent class A_λ, by Lemma 2.1, we observe that

$$A_\lambda \subset \bigcup_{x \in A_\lambda} B(x, \delta) \cap \Omega(f) \subset A_\lambda.$$

It implies that A_λ is open in $\Omega(f)$. Since $\Omega(f)$ is compact, we may write $\Omega(f) = \bigcup_{i=1}^{\ell} A_i$.

Let $r = \min\{d_H(A_i, A_j) : i \neq j \in \{1, \cdots, \ell\}\}$, where d_H denotes the Hausdorff distance. We take $0 < \alpha < \min\{\delta, r\}$. By Remark 2.2, we show that $f|_{\Omega(f)}$ is finite shadowing. Let $\xi = \{x_i\}_{i=0}^n$ be a finite α-pseudo orbit of f in $\Omega(f)$. It is clear that $\xi \subset A_i$ for some $i \in \{1, \cdots, \ell\}$. Since $x_0 \sim^\delta x_n$, there is a δ-chain $\{x_n, \cdots, x_m\}$ of f in A_i from x_n to x_0. We can extend the sequence $\{x_0, \cdots, x_m\}$ to a δ-pseudo orbit by letting $x_i = x_{i+km}$ for $0 \leq i \leq m - 1$ and $k \in \mathbb{Z}$. By the shadowing of f, there is $y \in X$ such that

$$d(f^i(y), x_i) \leq \varepsilon, \ \forall i \in \mathbb{Z}.$$

Since X is compact, there is a sequence $\{f^{k_j m}(y)\}_{j\in\mathbb{N}}$ such that $f^{k_j m}(y) \to z$ as $j \to \infty$ for some $z \in X$. We note that $z \in \Omega(f)$. Then, for $0 \le i \le n$, we have

$$d(f^i(z), x_i) = \lim_{j\in\infty} d(f^{i+k_j m}(z), x_{i+k_j m}) \le \varepsilon.$$

Then ξ is ε-shadowed by $z \in \Omega(f)$. This completes the proof.

2.2 Topological Stability

For given two homeomorphisms f and g on a compact metric space X, we define the distance

$$d_{C^0}(f, g) = \sup_{x\in X} d(f(x), g(x)).$$

In the following, we recall the notion of topologically stable homeomorphisms (see, for example, [39]).

Definition 2.10 A homeomorphism f on a compact metric space X is *topologically stable* if for every $\varepsilon > 0$ there is $\delta > 0$ such that for any homeomorphism $g : X \to X$ with $d_{C^0}(f, g) \le \delta$, there is a continuous map $h : X \to X$ such that $d_{C^0}(h, Id_X) \le \varepsilon$ and $f \circ h = h \circ g$.

Theorem 2.7 *Let f be a homeomorphism on a compact metric space X. If f is expansive and shadowing, then it is topologically stable.*

To prove it, we need the following two lemmas.

Lemma 2.2 *Let f be an expansive homeomorphism on a compact metric space X with an expansive constant $e > 0$. Then for any $\delta > 0$ there is $N \in \mathbb{N}$ such that if $x, y \in X$ satisfy $d(f^i(x), f^i(y)) \le e$ for all $-N \le i \le N$, then $d(x, y) \le \delta$.*

Proof. Let $f : X \to X$ be an expansive homeomorphism on a compact metric space with expansive constant $e > 0$. Suppose that there exists a constant $\delta > 0$ and for any $n \in \mathbb{N}$, there are $x_n, y_n \in X$ with $d(x_n, y_n) \ge \delta$ such that

$$d(f^i(x_n), f^i(y_n)) \le e, \quad \forall -n \le i \le n.$$

Since X is compact, we can assume that there are subsequences of $\{x_n\}_{n\in\mathbb{N}}$ and $\{y_n\}_{n\in\mathbb{N}}$, denoted by $\{x_{n_k}\}_{k\in\mathbb{N}}$ and $\{y_{n_k}\}_{k\in\mathbb{N}}$ respectively, such that $x_{n_k} \to x$ and $y_{n_k} \to y$ as $k \to \infty$. It follows easily that $d(x, y) \ge \delta > 0$, and so $x \ne y$.

On the other hand, for each $i \in \mathbb{Z}$, we have

$$d(f^i(x), f^i(y)) = \lim_{k\to\infty} d(f^i(x_{n_k}), f^i(y_{n_k})) \le e,$$

Since f is expansive, we get $x = y$. The contradiction completes the proof. □

Lemma 2.3 *Let f be an expansive, shadowing homeomorphism on a compact metric space X. Then for any $\varepsilon > 0$, there is $\delta > 0$ such that any δ-pseudo orbit of f can be ε-shadowed by a unique point in X.*

Proof. Let $e > 0$ be an expansive constant of f. For any $0 < \varepsilon < e/2$, we take $\delta > 0$ corresponding to $\varepsilon > 0$ by the shadowing of f. Let $\xi = \{x_i\}_{i\in\mathbb{Z}}$ be a δ-pseudo orbit of f. Suppose that ξ is ε-shadowed by two points $y, z \in X$. For each $i \in \mathbb{Z}$, we have

$$d(f^i(y), f^i(z)) \leq d(f^i(y), x_i) + d(x_i, f^i(z)) \leq 2\varepsilon < e.$$

By the expansiveness of f, we get $y = z$. This completes the proof.

□

Proof of Theorem 2.7. Let f be an expansive and shadowing homeomorphism on a compact metric space X. Let $\varepsilon > 0$ and $\varepsilon' = \frac{1}{4}\min\{\varepsilon, e\}$, where e is an expansive constant of f. By the shadowing of f, take $\delta > 0$ such that every δ-pseudo orbit of f is ε'-shadowed. Let $g : X \to X$ be a homeomorphisms on X with $d_{C^0}(f, g) \leq \delta$. For $x \in X$, we consider a sequence $\xi = \{g^i(x)\}_{i\in\mathbb{Z}}$ in X. Since

$$d(f(g^i(x)), g^{i+1}(x)) = d(f(g^i(x)), g(g^i(x)) \leq d(f, g) \leq \delta, \quad \forall i \in \mathbb{Z},$$

we see that ξ is a δ-pseudo orbit of f. By Lemma 2.3, ξ is ε'-shadowing by a unique point, denoted by $h(x)$, that is

$$d(f^i(h(x)), g^i(x)) \leq \varepsilon', \quad \forall i \in \mathbb{Z}.$$

Then we obtain the map $h : X \to X$ with $h(x) = y$ satisfying

$$d(f^i(h(x)), g^i(x)) \leq \varepsilon', \quad \forall x \in X \text{ and } \forall i \in \mathbb{Z}. \tag{2.1}$$

In particular, we have $d(h(x), x) \leq \varepsilon' < \varepsilon$ for all $x \in X$, and hence $d(h, Id_X) \leq \varepsilon$.

We now prove that $f \circ h(x) = h \circ g(x)$ for all $x \in X$. In fact, we have

$$d(f^i(h(g(x))), g^{i+1}(x)) \leq \varepsilon', \quad \forall x \in X \text{ and } \forall i \in \mathbb{Z}.$$

Moreover, by (2.1), we get

$$d(f^{i+1}(h(x)), g^{i+1}(x)) \leq \varepsilon'.$$

By Lemma 2.3, we get $f(h(x)) = h(g(x))$ for all $x \in X$.

Next, we show that h is continuous. Let $\varepsilon_1 > 0$. By Lemma 2.2, there is $N \in \mathbb{N}$ such that if $a, b \in X$ satisfy $d(f^i(a), f^i(b)) < e$ for all $-N \leq i \leq N$, then $d(a, b) < \varepsilon_1$. Since g is continuous, there is $\gamma > 0$ such that if $d(a, b) \leq \gamma$ for $a, b \in X$, then $d(g^i(a), g^i(b)) \leq \frac{e}{3}$ for every $-N \leq i \leq N$. Then for $x, y \in X$ with $d(x, y) \leq \gamma$, we have

$$
\begin{aligned}
d(f^i(h(x)), f^i(h(y)) &= d(h(g^i(x)), h(g^i(y))) \\
&\le d(h(g^i(x)), g^i(x)) + d(g^i(x), g^i(y)) + d(h(g^i(y)), g^i(y)) \\
&\le 2\varepsilon' + \frac{e}{3} \\
&< e, \quad \forall -N \le i \le N.
\end{aligned}
$$

Hence, by Lemma 2.2, we get $d(h(x), h(y)) \le \gamma$, and so h is continuous. This completes the proof.

□

2.3 Spectral Decomposition

In the last section of this chapter, we present another fundamental theorem in topological dynamics, which is known as the spectral decomposition theorem. We first give the definitions of topological transitivity and topological mixing as follows.

Definition 2.11 Let f be a homeomorphism on a compact metric space X.

(1) We say that f is *topologically transitive* on an invariant set $\Lambda \subset X$ if for any pair of nonempty open sets U and V in Λ, there is an integer $N > 0$ such that $f^N(U) \cap V \neq \emptyset$. f is said to be topologically transitive if it is topologically transitive on X.
(2) We say that f is *topologically mixing* on an invariant set $\Lambda \subset X$ if for any pair of nonempty open sets U and V in Λ, there is an integer $N > 0$ such that $f^n(U) \cap V \neq \emptyset$ for all $n \ge N$. f is said to be topologically mixing if it is topologically mixing on X.

Definition 2.12 We say that f admits the *spectral decomposition* if its nonwandering set $\Omega(f)$ is decomposed as the finite number of basic sets $\Omega_i \subset \Omega(f)$, i.e.,

$$\Omega(f) = \bigcup_{i=1}^{n} \Omega_i,$$

where Ω_i's are disjoint, closed and invariant, and f is topologically transitive on Ω_i for $1 \le i \le n$.

The classical spectral decomposition theorem was obtained by Smale [36] and Bowen [10]. More precisely, Smale [36] proved that if f is an Axiom A diffeomorphism on a compact smooth manifold then f admits the spectral decomposition. Afterwards, Bowen [10] showed that each basic set can be expressed as a finite disjoint union of closed sets C on which $f^m : C \to C$ is topologically mixing for some $m \in \mathbb{N}$.

There are many works that generalize the spectral decomposition theorem to more general settings (e.g., see [2, 16, 18, 29]). We know that if f is an Axiom A

diffeomorphism on a compact smooth manifold then f is expansive and shadowing on $\Omega(f)$. In this direction, Aoki [2] presents a topological version of the spectral decomposition theorem for expansive homeomorphisms with shadowing on compact metric spaces.

Theorem 2.8 *Let f be a homeomorphism on a compact metric space X. Suppose that f is expansive and shadowing on its chain recurrent set $CR(f)$. Then, $\Omega(f)$ decomposes as:*

(1) $\Omega(f) = \bigcup_{i=1}^{n} \Omega_i$, where Ω_i's are disjoint, closed and invariant, and f is topologically transitive on Ω_i for $1 \le i \le n$;
(2) for each $i \in \{1, \cdots, n\}$, there is a decomposition of Ω_i into disjoint closed sets $\Omega_i = \bigcup_{j=1}^{n_i} C_{j,i}$ such that $f(C_{j,i}) = C_{j+1,i}$ for $1 \le j < n_i$, $f(C_{n_i,i}) = C_{1,i}$, and the map $f^{n_i} : C_{j,i} \to C_{j,i}$ is topologically mixing for all $1 \le j \le n_i$.

To prove the theorem, we need some definitions and lemmas.

For any $x \in X$ and $\varepsilon > 0$, we define the *local stable* and *local unstable sets* W_ε^s and W_ε^u of x by

$$W_\varepsilon^s(x, f) = \{y \in X \mid d(f^n(x), f^n(y)) < \varepsilon, \ \forall n \in \mathbb{N}\},$$

$$W_\varepsilon^u(x, f) = \{y \in X \mid d(f^{-n}(x), f^{-n}(y)) < \varepsilon, \ \forall n \in \mathbb{N}\}.$$

The *stable* and *unstable sets* W^s and W^u of x in X are

$$W^s(x, f) = \{y \in X \mid d(f^n(x), f^n(y)) \to 0 \text{ as } n \to \infty\},$$

$$W^u(x, f) = \{y \in X \mid d(f^{-n}(x), f^{-n}(y)) \to 0 \text{ as } n \to \infty\}.$$

We observe that if $W^s(p, f) \cap W^u(q, f) \neq \emptyset$ and $W^u(p, f) \cap W^s(q, f) \neq \emptyset$ for some $p, q \in Per(f)$, then p and q are in a same chain component, i.e., $p \sim q$.

The following lemma shows the relationships between stable sets and local stable sets, and between unstable sets and local unstable sets under the assumption that f is expansive on $CR(f)$. It is useful in the proof of Theorem 2.8. We also note that the lemma also holds if f is expansive on $\Omega(f)$.

Lemma 2.4 *Let f be a homeomorphism on a compact metric space X. If f is expansive on $CR(f)$, there is a constant $\varepsilon > 0$ such that*

$$W_\varepsilon^s(x, f) \subset W^s(x, f) \text{ and } W_\varepsilon^u(x, f) \subset W^u(x, f), \ \forall x \in X.$$

Proof. Let $\varepsilon > 0$ be an expansive constant of f. We claim that $W_\varepsilon^s(x, f) \subset W^s(x, f)$ for any $x \in X$. Suppose, on the contrary, that there are $x \in X$ and $y \in W_\varepsilon^s(x, f) \setminus W^s(x, f)$. Then there are $r > 0$ and a sequence $\{n_k\}_{k \in \mathbb{N}}$ such that

$$r \le d(f^{n_k}(y), f^{n_k}(x)) \le \varepsilon \text{ for all } k \in \mathbb{N}.$$

By taking a subsequence, if necessary, we assume that $f^{n_k}(y)$ converges to $y_0 \in X$ and $f^{n_k}(x)$ converges to x_0 as $k \to \infty$. We note that $x_0, y_0 \in CR(f)$ and $x_0 \neq y_0$ since

$$d(x_0, y_0) = \lim_{k\to\infty} d(f^{n_k}(x), f^{n_k}(y)) \geq r.$$

Moreover, since $y \in W^s_\varepsilon(x, f)$, for $i \in \mathbb{Z}$ we have

$$d(f^i(x_0), f^i(y_0)) = \lim_{k\to\infty} d(f^i(f^{n_k}(x)), f^i(f^{n_k}(y))) \leq \varepsilon.$$

Since f is expansive on $CR(f)$, we obtain that $x_0 = y_0$. The contradiction shows that $W^s_\varepsilon(x, f) \subset W^s(x, f)$ for all $x \in X$. Similarly, we also have $W^u_\varepsilon(x, f) \subset W^u(x, f)$ for all $x \in X$. This completes the proof.

□

Lemma 2.5 *Let f be a homeomorphism on a compact metric space X. Suppose that f is expansive and shadowing on its chain recurrent set $CR(f)$. Then, $CR(f) = \overline{Per(f)}$.*

Proof. Suppose f is expansive and shadowing on its chain recurrent set $CR(f)$. Let $x \in CR(f)$, and let e be an expansive constant of f. Fix $0 < \varepsilon < \frac{e}{2}$. By the shadowing of f on $CR(f)$, we take a constant $\delta > 0$ such that every δ-pseudo orbit of f in $CR(f)$ is ε-shadowed by a point in X.

Since $x \in CR(f)$, by Remark 2.1, there is a δ-chain $\{x_i\}_{i=0}^m \subset CR(f)$ from x to itself. We can extend it to a δ-pseudo orbit $\{x_i\}_{i\in\mathbb{Z}}$ of f by letting $x_i = x_{i+km}$ for all $0 \leq i \leq m-1$ and $k \in \mathbb{Z}$. Since f is shadowing, there is $y \in X$ such that

$$d(f^i(y), x_i) \leq \varepsilon, \ \forall i \in \mathbb{Z}.$$

By Theorem 2.6, we note that the shadowing point y can be taken in $CR(f)$. For each $i \in \mathbb{Z}$, we have

$$\begin{aligned} d(f^i(y), f^i(f^m(y))) &\leq d(f^i(y), x_i') + d(x'_{m+i}, f^i(f^m(y))) \\ &\leq \varepsilon + \varepsilon \leq e. \end{aligned}$$

By the expansiveness of f on $CR(f)$, we get $f^m(y) = y$, and so y is a periodic point with $d(y, x) < \varepsilon$. Since ε is arbitrary, we obtain that $x \in \overline{Per(f)}$, and so $CR(f) \subset \overline{Per(f)}$. Therefore, $CR(f) = \overline{Per(f)}$, which completes the proof.

□

Proof of Theorem 2.8. Suppose that f is expansive and shadowing on its chain recurrent set $CR(f)$. By Lemma 2.5, we have

$$\Omega(f) = CR(f) = \bigcup_{\lambda\in\Lambda} \Omega_\lambda,$$

where Ω_λ's are chain components of f. By Lemma 2.4, let $\varepsilon_0 > 0$ be such that

$$W^s_{\varepsilon_0}(x, f) \subset W^s(x, f) \quad \text{and} \quad W^u_{\varepsilon_0}(x, f) \subset W^u(x, f) \quad \text{for all} \quad x \in X.$$

Since f is shadowing on $CR(f)$, by Theorem 2.6, we take $\delta > 0$ such that every δ-pseudo orbit of f in $CR(f)$ is ε_0-shadowed by a point in $CR(f)$. For fixed $\lambda \in \Lambda$, we will prove that Ω_λ is open in $\Omega(f)$. Let

$$U = \{y \in \Omega(f) \mid d(y, \Omega_\lambda) < \frac{\delta}{2}\}$$

be a neighborhood of Ω_λ in $\Omega(f)$. By Lemma 2.5, we have $\Omega(f) = \overline{Per(f)}$, and so $U \cap Per(f) \neq \emptyset$. Let $p \in U \cap Per(f)$, and $y \in \Omega_\lambda$ such that $d(y, p) < \delta/2$. We claim that $p \in \Omega_\lambda$. For any $0 < \tau < \delta/2$, let $p_\tau \in Per(f)$ be such that $d(p_\tau, y) < \tau$. By Lemma 2.1, we have $p_\tau \sim^\tau y$.

On the other hand, let $\xi_1 = \{z_i\}_{i \in \mathbb{Z}}$ be defined by $z_i = f^i(p)$ if $i < 0$ and $z_i = f^i(p_\tau)$ if $i \geq 0$. Since

$$d(f(f^{-1}(p)), p_\tau) \leq d(p, y) + d(y, p_\tau) < \frac{\delta}{2} + \tau < \delta,$$

we see that ξ_1 is a δ-pseudo orbit of f in $CR(f)$. By the shadowing of f, there is $z \in X$ such that

$$d(f^n(z), z_n) < \varepsilon_0 \text{ for all } n \in \mathbb{Z}.$$

Then $z \in W^u_{\varepsilon_0}(p, f) \cap W^s_{\varepsilon_0}(p_\tau, f) \subset W^u(p, f) \cap W^s(p_\tau, f)$. Let $k < 0 < l$ be such that $d(f(p), f^k(z)) < \tau$ and $d(f^l(z), p_\tau) < \tau$. Then the sequence $\{x_{k-1} = p, x_i = f^i(z) : k \leq i \leq l-1, x_l = p_\tau\}$ is a τ-chain of f from p to p_τ. Similarly, we can construct a τ-chain from p_τ to p, and so $p \sim^\tau p_\tau \sim^\tau y$. Since τ is arbitrary, we have $p \sim y$, which implies that $p \in \Omega_\lambda$. Consequently, we derive

$$\Omega_\lambda \supset \overline{U \cap Per(f)} \supset U \cap \overline{Per(f)} = U.$$

Then Ω_λ is open in $\Omega(f)$. Since $\Omega(f)$ is compact, $\Omega(f)$ can be expressed as a finite union of chain components of f, which are closed and invariant sets, that is, $\Omega(f) = \Omega_1 \cup \cdots \cup \Omega_n$.

Next, we will prove that f is topologically transitive on Ω_i for $1 \leq i \leq n$. Let U and V be two open sets in Ω_i for some $1 \leq i \leq n$. There is $0 < r < \delta/2$ such that

$$B(x, r) \cap \Omega_i \subset U \quad \text{and} \quad B(y, r) \cap \Omega_i \subset V$$

for some $x \in U$ and $y \in V$. By the shadowing of f on $CR(f)$, take $\delta_r > 0$ corresponding to r. Since $x, y \in \Omega_i$, there is a δ_r-chain $\{x_i\}_{i=0}^m$ of f from x to y. Since f is shadowing on $CR(f)$, there is $z \in CR(f)$ such that

$$d(z, x) < r < \frac{\delta}{2} \quad \text{and} \quad d(f^m(z), y) < r.$$

Then we have $z \in \Omega_i$ and

$$f^m(z) \in f^m(B(x,r)) \cap B(y,r) \cap \Omega_i \subset f^m(U) \cap V \neq \emptyset.$$

Therefore f is topologically transitive on Ω_i, and so it completes the proof of (1).

We now prove part (2). Fix $\Omega = \Omega_i$ for some $1 \le i \le n$. For each periodic point $p \in \Omega$, let $C_p = \overline{W^s(p, f) \cap \Omega}$ in Ω. Let $\varepsilon_1 = \min\{\frac{\varepsilon_0}{2}, \frac{\delta}{2}\}$. By the shadowing of f on $CR(f)$, take $\delta_1 > 0$ such that every δ_1-pseudo orbit of f in $CR(f)$ is ε_1-shadowed by a point in $CR(f)$.

Claim 1. C_p is open in Ω for any periodic point p in Ω.

Let m be the period of p, and $V = \{y \in \Omega \mid d(y, C_p) < \delta_1/4\}$ be a neighborhood of C_p in Ω. Since $\Omega \subset \overline{Per(f)}$, there is $q \in U \cap \Omega \cap Per(f)$ with period n. Let $x_0 \in W^s(p, f) \cap \Omega$ be such that $d(x_0, q) < \delta_1/2$. Consider a δ_1-pseudo orbit given by

$$y_i = \begin{cases} f^i(q) & \text{if } i < 0 \\ f^i(x_0) & \text{if } i \ge 0. \end{cases}$$

Since f is shadowing on $CR(f)$, there is $y \in \Omega(f)$ with $d(f^i(y), y_i) < \varepsilon_1$ for all $i \in \mathbb{Z}$. Then we derive the following:

- For $i = 0$, we have $d(y, x_0) < \delta/2$, and so $y \in U = \Omega$.
- For $i < 0$, we have $d(f^i(y), f^i(q)) < \varepsilon_0/2$, and so $y \in W^u_{\varepsilon_0}(q) \subset W^u(q, f)$.
- For $i \ge 0$, we have $d(f^i(y), f^i(x_0)) < \varepsilon_0/2$, and so $y \in W^s_{\varepsilon_0/2}(x_0)$.

Since $x_0 \in W^s(p, f)$, there is $N > 0$ such that if $i \ge N$, then $d(f^i(x_0), f^i(p)) \le \varepsilon_0/2$. Hence for any $i \in \mathbb{N}$ with $im > N$, and $j \ge 0$, we deduce that

$$\begin{aligned} d(f^j(f^{mi}(y)), f^j(p)) &= d(f^{j+mi}(y), f^{j+mi}(p)) \\ &\le d(f^{j+mi}(y), f^{j+mi}(x_0)) + d(f^{j+mi}(x_0), f^{j+mi}(p)) \le \varepsilon_0, \end{aligned}$$

and so,

$$f^{mi}(y) \in W^s_{\varepsilon_0}(p) \subset W^s(p, f) \text{ for any } i > N/m.$$

For $k \in \mathbb{N}$, since $f^{-mnk}(y) \in W^s(p, f)$ and $y \in \Omega$, we have

$$f^{-mnk}(y) \in \Omega \cap W^s(p, f) \subset C_p.$$

Since $y \in W^u(q, f)$, we get

$$d(f^{-mnk}(y), q) = d(f^{-mnk}(y), f^{-mnk}(q)) \to 0 \text{ as } k \to \infty.$$

Consequently, we have $q \in C_p$, and so $V \subset C_p$. This means that C_p is open in Ω.

Claim 2. If q is a periodic point with $q \in C_p$, then $C_p = C_q$.

We let m and n be the periods of p and q, respectively. We first show that $C_q \subset C_p$. Since $q \in C_p = \overline{W^s(p, f) \cap \Omega}$, there is a sequence $\{y_n\}_{n\in\mathbb{Z}}$ in $W^s(p, f) \cap \Omega$ such that $y_n \to q$ as $n \to \infty$. Fix $x_0 \in W^s(q, f) \cap \Omega$. For each $i \in \mathbb{N}$, let $\varepsilon_i = \dfrac{\varepsilon_0}{i+1}$ and take $\delta_i > 0$ such that every δ_i-pseudo orbit of f in $CR(f)$ is ε_i-shadowed by a point in X. Let $N_i \in \mathbb{N}$ be such that

$$d(y_{N_i}, q) < \delta_i/2 \qquad \text{and} \qquad d(f^{mnN_i}(x_0), q) < \delta_i/2.$$

We define a δ_i-pseudo orbit by

$$z_j = \begin{cases} f^j(x_0) & \text{if } j \le nmN_i - 1 \\ f^{j-nmN_i}(y_{N_i}) & \text{if } \ j \ge nmN_i. \end{cases}$$

Since f is shadowing on $CR(f)$, by Theorem 2.6, there is $x_i \in \Omega(f)$ such that $d(x_i, x_0) < \varepsilon_i$. Thus $x_i \to x_0$ as $i \to \infty$.

On the other hand, we have $x_i \in W^s(p, f)$ for all $i \in \mathbb{N}$. Indeed, for fixed $i \in \mathbb{N}$, we choose $M \in \mathbb{N}$ such that $f^M(p) = p$ and if $j \ge M$ then $d(f^j(y_{N_i}), f^j(p)) < \varepsilon_0/2$. Hence if $j \ge M$, we have

$$\begin{aligned} d\left(f^j(f^{nmN_i}(x_i)), f^j(p)\right) &\le d\left(f^{j+nmN_i}(x_i), f^j(y_{N_i})\right) + d\left(f^j(y_{N_i}), f^j(p)\right) \\ &\le \varepsilon_i + \varepsilon_0/2 \le \varepsilon_0. \end{aligned}$$

Then we obtain

$$f^{M+nmN_i}(x_i) \in W^s_{\varepsilon_0}(f^M(p)) = W^s_{\varepsilon_0}(p) \subset W^s(p, f).$$

Thus $x_i \in W^s(p, f) \cap \Omega$, and so $x_0 \in C_p$.

Next, we prove that $C_p \subset C_q$. By the above argument, it is sufficient to show that $p \in C_q$. Suppose not. Then we can take $d = \inf\{d(q, y) | y \in C_p \setminus C_q\} > 0$. Since $q \in C_p$, there is $x \in W^s(p, f) \cap \Omega$ such that $d(x, q) < d/2$. By definition of d, we note that $x \in C_q$. Then $f^{nmk}(x) \in C_q$ for all $k \in \mathbb{N}$. However, we have

$$d(f^{nmk}(x), p) = d(f^{nmk}(x), f^{nmk}(p)) \to 0 \text{ as } k \to \infty.$$

Since C_q is closed, we get $p \in C_q$, which is a contradiction. Hence $C_p = C_q$.

Claim 3. There is $a \in \mathbb{N}$ such that $\Omega = \bigcup_{i=0}^{a-1} C_{f^i(p)}$ and f^a is topologically mixing on C_p.

First, we show that there is $a \in \mathbb{N}$ such that $\Omega = \bigcup_{i=0}^{a-1} C_{f^i(p)}$. Let a be the smallest positive integer such that $C_{f^a(f^i(p))} = C_{f^i(p)}$ for all $i \in \mathbb{N}$. It is clear that $\bigcup_{i=0}^{a-1} C_{f^i(p)} \subset \Omega$. Suppose, on the contrary, that $\Omega \setminus \bigcup_{i=0}^{a-1} C_{f^i(p)}$ is nonempty. By the choice of $a > 0$, we observe that $\Omega \setminus \bigcup_{i=0}^{a-1} C_{f^i(p)}$ is invariant, and it is open in Ω by Claim 1. For two nonempty open sets C_p and $\Omega \setminus \bigcup_{i=0}^{a-1} C_{f^i(p)}$ in Ω, since f is topologically transitive on Ω, there is $m \in \mathbb{Z}$ such that

$$f^m(C_p) \cap \left(\Omega \setminus \bigcup_{i=0}^{a-1} C_{f^i(p)} \right) \neq \emptyset.$$

It contradicts the fact that $\Omega \setminus \bigcup_{i=0}^{a-1} C_{f^i(p)}$ is invariant, and so $\Omega = \bigcup_{i=0}^{a-1} C_{f^i(p)}$.

Next, we show that f^a is topologically mixing on C_p. Let U and V be any two nonempty open sets in C_p. Let q be a periodic point in V with the period n, and let $\eta > 0$ be such that $B(q, \eta) \cap C_p \subset V$. For $0 \leq j < n$, since U is open in $C_p = C_{f^{aj}(p)} = C_{f^{aj}(q)}$, there is $w_j \in U \cap W^s(f^{aj}(q), f)$. Then there is $M_j \in \mathbb{N}$ such that

$$d(f^i(w_j), f^i(f^{aj}(q))) < \eta \quad \text{for} \quad i \geq M_j.$$

Thus for $i > 0$ with $ani - aj \geq M_j$, we have

$$d(f^{ani-aj}(w_j), q) = d(f^{ani-aj}(w_j), f^{ani-aj}(f^{aj}(q))) < \eta.$$

Let $M \in \mathbb{N}$ be such that

$$M \geq \frac{1}{a} \max \left\{ M_j | 0 \leq j < n \right\}.$$

For $k \geq M$, we have $ak = a(ni - j) \geq M_j$ for some $i, j \in \mathbb{N}$ and $0 \leq j < n$. Then

$$d(f^{ak}(w_j), q) = d(f^{ani-aj}(w_j), q) < \eta.$$

Therefore, $f^{ka}(U) \cap V \neq \emptyset$ for all $k \geq M$, and so f^a is topologically mixing on C_p. □

We note that if f is expansive and shadowing on its nonwandering set $\Omega(f)$, then we cannot guarantee the existence of the spectral decomposition of f by the forthcoming Remark 2.3. For this, we first prove the following proposition.

Proposition 2.1 *Let f be a homeomorphism on a compact metric space X which is expansive and shadowing on its nonwandering set $\Omega(f)$. If $\Omega(f)$ admits the spectral decomposition, then $\Omega(f) = \overline{Per(f)}$.*

Proof. Suppose the nonwandering set $\Omega(f)$ is decomposed as a finite union of disjoint closed invariant sets Ω_i, $\Omega(f) = \bigcup_{i=1}^{l} \Omega_i$, such that f is topologically transitive on Ω_i for each $1 \leq i \leq l$. Take a small constant $r > 0$ such that the r-neighborhoods of Ω_i $(i = 1, \ldots, l)$ are pairwise disjoint. Let $e > 0$ be an expansive constant of f on $\Omega(f)$, and let $0 < \varepsilon < \min\{r, e/2\}$. Since f is shadowing on $\Omega(f)$, we take a constant $\delta > 0$ such that every δ-pseudo orbit of f in $\Omega(f)$ is ε-shadowed by a point in X. Let $x \in \Omega(f)$. Then there exists Ω_j $(1 \leq j \leq l)$ containing x. Since f is topologically transitive on Ω_j, we can construct a finite δ-pseudo orbit $\{x_i\}_{i=0}^{n-1}$ in Ω_j from x to itself. We extend it to a δ-pseudo orbit $\xi = \{x_i\}_{i \in \mathbb{Z}}$ by defining $x_i = x_{i+kn}$ for all $k \in \mathbb{Z}$ and $0 \leq i \leq n-1$. Since f is shadowing on $\Omega(f)$, there is a point $y \in X$ such that

$$d(f^i(y), x_i) \leq \varepsilon, \text{ for all } i \in \mathbb{Z}.$$

Since X is compact, we take a subsequence $\{k_j\}_{j\in\mathbb{N}}$ in $\mathbb{N}$ such that $f^{k_jn}(y) \to p$ as $j \to \infty$ for some $p \in \Omega(f)$. We note that $p \in \Omega_j$ since

$$d(x, p) = \lim_{j\to\infty} d(x_{k_jn}, f^{k_jn}(y)) \leq \varepsilon < r.$$

For any $i \in \mathbb{Z}$, we have

$$d(f^i(p), x_i) = \lim_{j\to\infty} d(f^{i+k_jn}(y), x_{i+k_jn}) \leq \varepsilon.$$

Moreover, since

$$d(f^i(f^n(p)), f^i(p)) \leq d(f^i(f^n(p)), x_{i+n}) + d(x_i, f^i(p)) < e$$

for all $i \in \mathbb{Z}$, by expansiveness of f on $\Omega(f)$, we see that $f^n(p) = p$, and so p is a periodic point of f. Since ε is arbitrary, we see that $x \in \overline{Per(f)}$. This completes the proof.

□

We observe that the converse of Proposition 2.1 also holds. Precisely, if f is expansive and shadowing on $\Omega(f) = \overline{Per(f)}$, then f admits the spectral decomposition. The proof can be derived by repeating the argument used in the proof of Theorem 2.8 (1).

A natural question arising from Theorem 2.8 is if we can replace $CR(f)$ in its statement by other sets like $\Omega(f)$. The answer is negative by the remark below.

Remark 2.3 There is a compact metric space X and a homeomorphism $f : X \to X$ which is expansive and shadowing in $\Omega(f)$ but f does not admit the spectral decomposition. In fact, Dankner constructed a diffeomorphism f on a compact smooth manifold X with $\dim X \geq 3$ such that the nonwandering set $\Omega(f)$ is hyperbolic for f but $\Omega(f) \neq \overline{Per(f)}$ (for more details, see Theorem 6.1 in [17]). Since $\Omega(f)$ is hyperbolic, f is expansive and shadowing on $\Omega(f)$. Then, f does not admit the spectral decomposition by Proposition 2.1.

Chapter 3
Expansive Measures

In this chapter, we study the notion of expansive measures of a homeomorphism f on a compact metric space X. In Sect. 3.1, we introduce a notion of an expansive measure of f, and present its characterizations by using the isolated diagonal of $X \times X$ and generators. In Sect. 3.2, we show that a circle homeomorphism admits an expansive measure if and only if it is a Denjoy homeomorphism. In Sect. 3.3, we present a class of examples for expansive invariant measures for homeomorphisms with positive entropy. In Sect. 3.4, we characterize measure expansive homeomorphisms by using the notion of countable expansiveness, and study the relationship between their local stable sets and stable sets.

3.1 Expansive Measures for Homeomorphisms

Let f be a homeomorphism on a compact metric space X, and $\mu \in \mathcal{M}(X)$. For given $x \in X$ and $\delta > 0$, we write

$$\Gamma_\delta^f(x) = \{y \in X : d(f^i(x), f^i(y)) \leq \delta, \ \forall i \in \mathbb{Z}\}.$$

The set $\Gamma_\delta^f(x)$ is called the *dynamic δ-ball* of f centered at x. We note that $\Gamma_\delta^f(x)$ is compact. Using this notation, we can reformulate the notion of expansiveness as follows: a homeomorphism f is expansive if and only if there is a constant $\delta > 0$ such that $\Gamma_\delta^f(x) = \{x\}$ for all $x \in X$.

Definition 3.1 A Borel measure $\mu \in \mathcal{M}(X)$ is said to be *expansive* for f (or f is μ-expansive) if there is a constant $\delta > 0$ such that $\mu(\Gamma_\delta^f(x)) = 0$ for all $x \in X$. The constant δ is called an *expansive constant* of μ.

We observe that any expansive measure of a homeomorphism f is nonatomic. In fact, let μ be an expansive measure of f, and $\delta > 0$ be an expansive constant of μ.

K. Lee et al., *Topological Dynamics from a Measure-Theoretic Viewpoint*, Springer Asia Pacific Mathematics Series 14, https://doi.org/10.1007/978-981-95-6390-6_3

Suppose that μ is atomic. Then there exists $x \in X$ such that $\mu(\Gamma_\delta^f(x)) \geq \mu(\{x\}) > 0$. The contradiction shows that μ is nonatomic.

Moreover, we see that if f is expansive, then every nonatomic measure on X is expansive for f.

Remark 3.1 Let f and g be homeomorphisms on compact metric spaces X and Y, respectively, which are conjugate, i.e., there is a homeomorphism $h : X \to Y$ such that $g = h \circ f \circ h^{-1}$. If $\mu \in \mathcal{M}(X)$ is expansive for f, then $h_*(\mu)$ is expansive for g.

In fact, let $\delta > 0$ be an expansive constant of μ. By the continuity of h, there is $\varepsilon > 0$ such that

$$\Gamma_\varepsilon^g(x) \subset h(\Gamma_\delta^f(h^{-1}(x))), \ \forall x \in X.$$

Then, for any $y \in Y$, we have $h_*(\mu)(\Gamma_\varepsilon^g(x)) \leq \mu(\Gamma_\delta^f(h^{-1}(x))) = 0$. Hence $h_*(\mu)$ is expansive for g.

Moreover, we observe that expansiveness for measures does not depend on the choice of metrics on X. Let d and D be equivalent metrics on X, and suppose that μ is expansive for f on (X, d). We see that the map $h : (X, d) \to (X, D)$ given by $h(x) = x$ for all $x \in X$ is a homeomorphism and satisfies $f = h \circ f \circ h^{-1}$. Hence $\mu \in \mathcal{M}(X)$ is expansive for f on (X, D).

Definition 3.2 A homeomorphism f on X is said to be *equicontinuous* if for any $\varepsilon > 0$, there is $\delta > 0$ such that if $x, y \in X$ satisfy $d(x, y) < \delta$ then

$$d(f^n(x), f^n(y)) < \varepsilon, \ \forall n \in \mathbb{Z}.$$

Moreover, we say that a f is an *isometry* under the metric d if $d(f(x), f(y)) = d(x, y)$ for all $x, y \in X$.

In the following example, we show that every equicontinuous homeomorphism does not have an expansive measure.

Example 3.1 Let f be an equicontinuous homeomorphism on a compact metric space X. Then f does not have an expansive measure. Indeed, suppose, on the contrary, that there is an expansive measure μ for f. We define a metric $D : X \times X \to \mathbb{R}$ by

$$D(x, y) = \sup\{d(f^n(x), f^n(y)) : n \in \mathbb{Z}\}, \ \forall x, y \in X.$$

By the equicontinuity of f, the metric D induces the topology on X. Moreover, f is an isometry under the metric D. By Remark 3.1, μ is expansive for f under D. Let $\delta > 0$ be an expansive constant of μ. Since μ is expansive for isometry f, we have $\Gamma_\delta^f(x) = B[x, \delta]$ for all $x \in X$, where $B[x, \delta]$ denotes the closed δ-ball centered at x. Since X is compact, there is a finite sequence $\{x_1, \cdots, x_n\}$ in X such that $X = \bigcup_{i=1}^n B(x_i, \delta)$. Then we have

$$1 = \mu(X) \leq \sum_{i=1}^n \mu(B(x_i, \delta)) \leq \sum_{i=1}^n \mu(\Gamma_\delta^f(x_i)) = 0.$$

The contradiction completes the proof.

□

Next, we give two characterizations of expansive measures for f by using generator and diagonal of $X \times X$. For this, we need the following lemma.

Lemma 3.1 *Let f be a homeomorphism on a compact metric space X, and $\mu \in \mathcal{M}(X)$. Then μ is expansive for f if and only if there are a constant $\delta > 0$ and a Borel set $B \in \beta_X$ with $\mu(B) = 1$ such that $\mu(\Gamma^f_\delta(x)) = 0$ for all $x \in B$.*

Proof. It is enough to prove the if part. Suppose there are a constant $\delta > 0$ and a Borel set $B \in \beta_X$ with $\mu(B) = 1$ such that $\mu(\Gamma^f_\delta(x)) = 0$ for all $x \in B$. We claim that μ is expansive for f with an expansive constant $\delta/2$. Suppose, on the contrary, that there is $x_0 \in X$ such that $\mu(\Gamma^f_{\delta/2}(x_0)) > 0$. We let $B_0 = \{x \in X : \mu(\Gamma^f_\delta(x)) = 0\}$. Since $B \subset B_0$, we have $\mu(B_0) = 1$. Moreover, since μ is a probability measure, there exists $y_0 \in B_0 \cap \Gamma^f_{\delta/2}(x_0)$.

Let $x \in \Gamma^f_{\delta/2}(x_0)$. Since

$$\begin{aligned} d(f^i(x), f^i(y_0)) &\le d(f^i(x), f^i(x_0)) + d(f^i(x_0), f^i(y_0)) \\ &\le \delta \text{ for all } i \in \mathbb{Z}, \end{aligned}$$

we see that $x \in \Gamma^f_\delta(y_0)$, and so $\Gamma^f_{\delta/2}(x_0) \subset \Gamma^f_\delta(y_0)$. Since $y_0 \in B_0$, we get $\mu(\Gamma^f_{\delta/2}(x_0)) \le \mu(\Gamma^f_\delta(y_0)) = 0$. The contradiction completes the proof.

□

For a compact metric space X and a homeomorphism on X, we define a metric D on $X \times X$ by

$$D((x, y), (z, w)) = d(x, z) + d(y, w), \quad \text{for } (x, y), (z, w) \in X \times X,$$

and a map $f \times f : X \times X \to X \times X$ by

$$(f \times f)(x, y) = (f(x), f(y)), \quad \text{for } (x, y) \in X \times X.$$

It is clear that $f \times f$ is a homeomorphism on $X \times X$. We denote by $\Delta = \{(x, x) : x \in X\}$ the diagonal of $X \times X$.

We say that Δ is an *isolated set* of $f \times f$ if there exists a compact neighborhood U of Δ in $X \times X$ such that

$$\Delta = \{z \in U : (f \times f)^n(z) \in U, \ \forall n \in \mathbb{Z}\} = \bigcap_{n \in \mathbb{Z}} (f \times f)^n(U).$$

It is known that f is expansive if and only if the diagonal Δ is an isolated set of $f \times f$ (for example see [1]). In the following theorem, we present a measurable version of the fact. For this, we introduce the notion of an isolated subset with respect to a Borel measure.

Definition 3.3 An f-invariant set $\Delta \subset X$ is said to be *μ-isolated* for f if there exists a compact neighborhood U of Δ such that

$$\mu(\{z \in X : f^n(z) \in U, \ \forall n \in \mathbb{Z}\}) = 0.$$

For simplicity, we write $\mu^2 = \mu \times \mu$ as the product measure on $X \times X$.

Theorem 3.1 *Let f be a homeomorphism on a compact metric space X, and $\mu \in \mathcal{M}(X)$. Then $\mu \in \mathcal{M}(X)$ is expansive for f if and only if the diagonal Δ is μ^2-isolated for $f \times f$.*

Proof. For $\delta > 0$, we denote by $U_\delta = \{z \in X \times X : d(z, \Delta) \le \delta\}$ the δ-neighborhood of Δ, and $g = f \times f$.

We first claim that

$$\{z \in X \times X : g^n(z) \in U_\delta, \ \forall n \in \mathbb{Z}\} = \bigcup_{x \in X} (\{x\} \times \Gamma_\delta^f(x)).$$

In fact, let $z = (x, y) \in X \times X$ be such that $g^n(z) \in U_\delta$ for all $n \in \mathbb{Z}$. Then for $n \in \mathbb{Z}$, there is $p_n \in X$ such that

$$D(g^n(x, y), (p_n, p_n)) = d(f^n(x), p_n) + d(f^n(y), p_n) \le \delta,$$

and so $d(f^n(x), f^n(y)) \le \delta$. Hence we have $y \in \Gamma_\delta^f(x)$, and so $z \in \bigcup_{x \in X}\{x\} \times \Gamma_\delta^f(x)$.

Conversely, let $z = (x, y) \in X \times X$ be such that $y \in \Gamma_\delta^f(x)$. Then we have $d(f^n(x), f^n(y)) \le \delta$ for all $n \in \mathbb{Z}$. It means that

$$D(g^n(x, y), (f^n(x), f^n(x))) = d(f^n(x), f^n(y)) \le \delta$$

for all $n \in \mathbb{Z}$. This implies that $g^n(z) \in U_\delta$ for all $n \in \mathbb{Z}$. The claim is proved.

For $\delta > 0$, we denote by F_δ the characteristic map given by $F(x, y) = \chi_{\Gamma_\delta^f(x)}(y)$ for all $(x, y) \in X \times X$. Then we have

$$\mu^2(\{z \in X \times X : g^n(z) \in U_\delta, \ \forall n \in \mathbb{Z}\}) = \int_X \int_X \chi_{\Gamma_\delta^f(x)}(y) d\mu(y) d\mu(x). \quad (3.1)$$

Suppose that $\mu \in \mathcal{M}(X)$ is an expansive measure of f with an expansive constant δ. Then we have

$$\int_X \chi_{\Gamma_\delta^f(x)}(y) d\mu(y) = 0, \qquad \forall x \in X.$$

Hence, by (3.1), we get $\mu^2(\{z \in X \times X : g^n(z) \in U_\delta, \ \forall n \in \mathbb{Z}\}) = 0$.

Conversely, suppose that the diagonal Δ is μ^2-isolated for $f \times f$, i.e., there is a constant $\delta > 0$ such that $\mu^2(\{z \in X \times X : g^n(z) \in U_\delta, \ \forall n \in \mathbb{Z}\}) = 0$. By (3.1), we obtain that $\mu(\Gamma^f_\delta(x)) = 0$ for μ-almost every $x \in X$, and so μ is expansive for f by Lemma 3.1. This completes the proof.

□

A finite open covering $\mathcal{A}$ of X is called a *generator* of f if for every bisequence $\{A_n \in \mathcal{A}\}_{n\in\mathbb{Z}}$,

$$\bigcap_{n\in\mathbb{Z}} f^{-n}(\overline{A_n}) \text{ has at most one point.}$$

It is well known that f is expansive if and only if it has a generator. Using this idea, we introduce a measurable version of generators as follows.

Definition 3.4 Let f be a homeomorphism on a compact metric space X, and let $\mu \in \mathcal{M}(X)$. We say that a finite open covering $\mathcal{A}$ of X is a *μ-generator* of f if for every sequence $\{A_n \in \mathcal{A}\}_{n\in\mathbb{Z}}$, one has

$$\mu\left(\bigcap_{n\in\mathbb{Z}} f^{-n}(\overline{A_n})\right) = 0.$$

Theorem 3.2 *Let f be a homeomorphism on a compact metric space X, and $\mu \in \mathcal{M}(X)$. Then μ is expansive for f if and only if f has a μ-generator.*

Proof. Suppose that μ is expansive for f with an expansive constant $\delta > 0$. We claim that there is a μ-generator of f. In fact, we take $\mathcal{A}$ as the collection of all open δ-balls $B(x, \delta)$ for $x \in X$. Then for any sequence $\{A_n\}_{n\in\mathbb{Z}}$ in $\mathcal{A}$, we have

$$\bigcap_{n\in\mathbb{Z}} f^{-n}(\overline{A_n}) \subset \Gamma^f_\delta(x), \quad \forall x \in \bigcap_{n\in\mathbb{Z}} f^{-n}(\overline{A_n}),$$

and so

$$\mu\left(\bigcap_{n\in\mathbb{Z}} f^{-n}(\overline{A_n})\right) \leq \mu(\Gamma^f_\delta(x)) = 0.$$

Hence $\mathcal{A}$ is a μ-generator of f.

Conversely, suppose that f has a μ-generator $\mathcal{A}$. Let $\delta > 0$ be a Lebesgue number of $\mathcal{A}$. Let $x \in X$. For every $n \in \mathbb{Z}$ there is $A_n \in \mathcal{A}$ such that the closed δ-ball centered at $f^n(x)$ belongs to $\overline{A_n}$. Then we have

$$\Gamma^f_\delta(x) \subset \bigcap_{n\in\mathbb{Z}} f^{-n}(\overline{A_n}).$$

Since $\mathcal{A}$ is a μ-generator of f, we get $\mu(\Gamma^f_\delta(x)) = 0$. Since x is arbitrary, μ is expansive for f. This completes the proof.

□

3.2 Expansive Measures on the Circle

In this section, we study expansive measures for circle homeomorphisms. Let f be a homeomorphism on the circle S^1. We define a projection $\pi : \mathbb{R} \to S^1$ by $\pi(x) = x$ mod $\mathbb{Z}$. For any orientation-preserving homeomorphism f on the circle S^1, there exists a homeomorphism $F : \mathbb{R} \to \mathbb{R}$ of f satisfying

$$\pi \circ F = f \circ \pi \text{ and } F(x+m) = F(x) + m, \ \forall x \in \mathbb{R} \text{ and } m \in \mathbb{Z}.$$

The function F is unique modulo integer, and it is called a *lift* of f.

The *rotation number* of f is given by

$$r(f) = \lim_{n \to \infty} \frac{F^n(x) - x}{n}$$

We remark that $r(f)$ exists and is independent of the choice of x (mod $\mathbb{Z}$).

Definition 3.5 An orientation-preserving homeomorphism f on the circle is called a *Denjoy homeomorphism* if it has an irrational rotation number and is not conjugate to a rotation.

We show that f admits an expansive measure if and only if it is a Denjoy homeomorphism. For this, we need some notations and lemmas. Let f be a homeomorphism on a compact metric space X, and $\mu \in \mathcal{M}(X)$.

Definition 3.6 We say that $x \in X$ is a *point with converging semiorbits* under f if both $\alpha(x, f)$ and $\omega(x, f)$ are singleton. Denote by $A(f)$ the set of points with converging semiorbits under f.

For $x, y \in X$, $n \in \mathbb{N}$ and $m \in \mathbb{N}$, we define

$$A(x, y, n, m) = \left\{ z \in X : \max\{d(f^{-i}(z), x), d(f^i(z), y)\} \le \frac{1}{n}, \quad \forall i \ge m \right\}.$$

In the following lemma, we show the relationship between two sets $A(f)$ and $A(x, y, n, m)$.

Lemma 3.2 *Let f be a homeomorphism on a compact metric space X. Then there is a sequence $\{x_k\}_{k\in\mathbb{N}}$ in X such that*

$$A(f) \subset \bigcap_{n\in\mathbb{N}} \bigcup_{k,\ell,m\in\mathbb{N}} A(x_k, x_\ell, n, m). \tag{3.2}$$

Proof. Since X is separable, it admits a dense subset $A = \{x_k\}_{k\in\mathbb{N}}$. Let $z \in A(f)$ and $n \in \mathbb{N}$. Then there are two fixed points $x, y \in X$ of f such that $\alpha(x, f) = \{x\}$ and $\omega(x, f) = \{y\}$. Hence we can take $m \in \mathbb{N}$ such that

$$d(f^{-i}(z), x) \le \frac{1}{2n} \text{ and } d(f^{i}(z), y) \le \frac{1}{2n}, \quad \forall i \ge m.$$

Since A is dense in X, there are $x_k, x_\ell \in A$ such that

$$d(x, x_k) \le \frac{1}{2n} \text{ and } d(y, x_{k'}) \le \frac{1}{2n}.$$

Then we have

$$d(f^{-i}(z), x_k) \le d(f^{-i}(z), x) + d(x, x_k) \le \frac{1}{2n} + \frac{1}{2n} = \frac{1}{n},$$

and

$$d(f^{i}(z), x_\ell) \le d(f^{i}(z), y) + d(y, x_\ell) \le \frac{1}{2n} + \frac{1}{2n} = \frac{1}{n},$$

for all $i \ge m$. It means that $z \in A(x_k, x_\ell, n, m)$, and so it completes the proof.

□

Lemma 3.3 *Let f be a homeomorphism on a compact metric space X, and $\mu \in \mathcal{M}(X)$. If μ is expansive for f, then $\mu(A(f)) = 0$.*

Proof. Suppose $\mu \in \mathcal{M}(X)$ is expansive for f. Assume, on the contrary, that $\mu(A(f)) > 0$. Let $\{x_k\}_{k\in\mathbb{N}}$ be a sequence as in Lemma 3.2. Then by (3.2) we have

$$\mu\left(\bigcup_{k,\ell,m\in\mathbb{N}} A(x_k, x_\ell, n, m)\right) > 0 \qquad \forall n \in \mathbb{N}.$$

Let $e > 0$ be an expansive constant of μ and take $n \ge \frac{2}{e}$. Then there are $k, \ell, m \in \mathbb{N}$ such that

$$\mu(A(x_k, x_\ell, n, m)) > 0.$$

We now show that there are $z \in A(x_k, x_\ell, n, m)$ and a constant $\delta_0 > 0$ such that

$$\mu(A(x_k, x_{k'}, n, m) \cap B[z, \delta]) > 0 \text{ for all } 0 < \delta < \delta_0,$$

where $B[z, \delta]$ denotes the closed δ-ball centered at z. Suppose, on the contrary, that for every $z \in A(x_k, x_\ell, n, m)$ we could find $\delta_z > 0$ such that

$$\mu(A(x_k, x_{k'}, n, m) \cap B[z, \delta_z]) = 0.$$

We observe that

$$\xi = \left\{B\left(z, \frac{\delta_z}{2}\right) : z \in A(x_k, x_{k'}, n, m)\right\}$$

is an open covering of $A(x_k, x_\ell, n, m)$. Since X is a separable metric space, $A(x_k, x_{k'}, n, m)$ is also separable, and so it is Lindelöf. Then ξ has a countable subcover $\{B_i\}_{i\in\mathbb{N}}$ of $A(x_k, x_{k'}, n, m)$. Hence we get

$$\mu(A(x_k, x_{k'}, n, m)) \leq \sum_{i\in\mathbb{N}} \mu(A(x_k, x_{k'}, n, m) \cap B_i) = 0.$$

The contradiction shows that there exist $z \in A(x_k, x_\ell, n, m)$ and a constant $\delta_0 > 0$ such that

$$\mu(A(x_k, x_{k'}, n, m) \cap B[z, \delta]) > 0 \text{ for all } 0 < \delta < \delta_0.$$

On the other hand, since f is continuous, there is a constant $0 < \delta_1 < \delta_0$ such that if $d(z, w) < \delta_1$ then

$$d(f^i(z), f^i(w)) < \frac{e}{2} \text{ for all } -m \leq i \leq m. \tag{3.3}$$

We claim that

$$A(x_k, x_{k'}, n, m) \cap B[z, \delta_1] \subset \Gamma_e^f(z).$$

Let $w \in A(x_k, x_{k'}, n, m) \cap B[z, \delta_1]$. By (3.3), we have

$$d(f^i(w), f^i(z)) \leq e, \quad \forall -m \leq i \leq m.$$

Moreover, since $z, w \in A(x_k, x_\ell, n, m)$, we get

$$d(f^{-i}(w), f^{-i}(z)) \leq d(f^{-i}(w), x_k) + d(f^{-i}(z), x_k) \leq \frac{1}{n} + \frac{1}{2} \leq e,$$

and

$$d(f^i(w), f^i(z)) \leq d(f^i(w), x_{k'}) + d(f^i(z), x_\ell) \leq e, \quad \forall i \geq m.$$

It implies that $w \in \Gamma_e^f(z)$, and so

$$A(x_k, x_\ell, n, m) \cap B[z, \delta_1] \subset \Gamma_e^f(z).$$

Therefore we see that

$$0 < \mu(A(x_k, x_{k'}, n, m) \cap B[z, \delta_1]) \leq \mu(\Gamma_e^f(z)) = 0$$

which is a contradiction. This completes the proof. □

Corollary 3.1 *Let f be a homeomorphism on a compact metric space X. If f is expansive, then $A(f)$ is countable.*

Proof. Suppose that $A(f)$ is uncountable. By Theorem 1.6, there exists a nonatomic measure μ on X such that $\mu(A(f)) = 1$. Since f is expansive, μ is expansive for f. It contradicts Lemma 3.3 which says that $\mu(A(f)) = 0$. □

Lemma 3.4 *Let f be a homeomorphism on a compact metric space X. Suppose $\omega(x, f)$ is finite for some $x \in X$. Then there is a periodic point $z \in Per(f)$ such that $d(f^n(x), f^n(z)) \to 0$ as $n \to \infty$.*

Proof. Let F be a proper invariant closed subset of $\omega(x, f)$. We first claim that $F \cap \overline{(\omega(x, f) \setminus F)} \neq \emptyset$. Suppose that there are two open sets U_1 and U_2 such that $\omega(x, f) \setminus F \subset U_1$, $F \subset U_2$ and $\overline{U_2} \cap f(\overline{U_1}) = \emptyset$. By the definition of $\omega(x, f)$, there are infinitely many n's such that $f^n(x)$ belongs to U_1 or U_2. Then there is a sequence $\{n_k\}_{k \in \mathbb{N}}$ such that $f^{n_k}(x) \in U_1$ and $f^{n_{k+1}}(x) \in U_2$. Let y be a limit point of $\{f^{n_k}(x)\}_{k \in \mathbb{N}}$. It is easy to see that $y \in \overline{U_1} \cap f^{-1}(\overline{U_2})$, and so $f(\overline{U_1}) \cap \overline{U_2} \neq \emptyset$. The contradiction shows that $F \cap \overline{(\omega(x, f) \setminus F)} \neq \emptyset$.

Now since $\omega(x, f)$ is finite, there is a periodic orbit $P \subset \omega(x, f)$. If $P \neq \omega(x, f)$, by letting $F = \omega(x, f) \setminus P$ and applying the claim, we see that

$$\emptyset \neq F \cap \overline{(\omega(x, f) \setminus F)} = (\omega(x, f) \setminus P) \cap P = \emptyset.$$

The contradiction shows that $P = \omega(x, f)$. Then the result is easily followed. □

Definition 3.7 A point $x \in X$ is said to be *heteroclinic* for f if both $\alpha(x, f)$ and $\omega(x, f)$ are periodic points. Denote by $Het(f)$ the set of all heteroclinic points of f.

It is clear that $A(f) \subset Het(f)$. We are now going to show that the set of heteroclinic points of a homemorphism on a compact metric space has measure zero with respect to any expansive measure, which improves Lemma 3.3.

Lemma 3.5 *Let f be a homeomorphism on a compact metric space X. Then*

$$Het(f) \subset \bigcup_{n \in \mathbb{N}} A(f^n).$$

Proof. Let $x \in Het(f)$. By definition, we see that $\alpha(x, f)$ and $\omega(x, f)$ are finite sets. By Lemma 3.4, there is a periodic point $y \in Per(f)$ such that $d(f^n(x), f^n(y)) \to 0$ as $n \to \infty$. We denote by n_y the period of y. Then we have $d(f^{kn_y}(x), y) \to 0$ as $k \to \infty$, and so $\omega(x, f^{n_y}) = \{y\}$. Similarly, there is a periodic point z with period n_z such that $\alpha(x, f^{n_z}) = \{z\}$. By taking $m = n_y n_z$, we observe that $\alpha(x, f^n) = \{z\}$ and $\omega(x, f^n) = \{y\}$. It follows that $x \in A(f^n)$, which completes the proof. □

Now let us give a characterization of dynamic ball as follows. For given $m \in \mathbb{N}$, $x \in X$ and $\delta > 0$, we define

$$V_f[x, m, \delta] = \{y \in X : d(f^i(x), f^i(y)) \leq \delta, \ \forall -m \leq i \leq m\}.$$

It is clear that $V[x, m, \delta] \subset V[x, n, \delta]$ for $n \leq m$, and

$$\Gamma_\delta^f(x) = \bigcap_{n \in \mathbb{N}} V[x, n, \delta].$$

For a Borel measure $\mu \in \mathcal{M}(X)$, we see that

$$\mu(\Gamma_\delta^f(x)) = \lim_{m \to \infty} \mu(V[x, m, \delta]).$$

Hence, a Borel measure μ on X is expansive for f if and only if there is a constant $\delta > 0$ such that

$$\liminf_{m \to \infty} \mu(V[x, m, \delta]) = 0, \ \forall x \in X.$$

Lemma 3.6 *Let f be a homeomorphism on a compact metric space X, and $\mu \in \mathcal{M}(X)$. Then μ is expansive for f if and only if it is expansive for f^n for all $n \in \mathbb{Z} \setminus \{0\}$.*

Proof. Without loss of generality, we assume that $n > 0$. Suppose that $\mu \in \mathcal{M}(X)$ is an expansive measure of f with an expansive constant $\delta > 0$. Since X is compact, there is a constant $0 < \varepsilon < \delta$ such that if $d(x, y) \leq \varepsilon$, then

$$d(f^i(x), f^i(y)) < \delta \text{ for all } -n \leq i \leq n.$$

Then we see that $\Gamma_\varepsilon^{f^n}(x) \subset \Gamma_\delta^f(x)$, and so $\mu(\Gamma_\varepsilon^{f^n}(x)) \leq \mu(\Gamma_\delta^f(x)) = 0$ for all $x \in X$. Hence μ is expansive for f^n.

Conversely, suppose that μ is an expansive measure of f^n. For $\delta > 0$, we observe that

$$V_f[x, n \cdot m, \delta] \subset V_{f^n}[x, m, \delta].$$

Since μ is expansive for f^n, there is a constant $\delta > 0$ such that for every $x \in X$ there is a sequence $\{m_j\}_{j \in \mathbb{N}}$ such that $m_j \to \infty$ and

$$\mu(V_{f^n}[x, m_j, \delta]) \to 0 \text{ as } j \to \infty.$$

This implies that

$$\liminf_{n \to \infty} \mu(V_f[x, n, \delta]) = 0.$$

Since x is arbitrary, we derive that μ is expansive for f. This completes the proof. □

Proposition 3.1 Let f be a homeomorphism on a compact metric space X. If a Borel measure $\mu \in \mathcal{M}(X)$ is expansive for f, then $\mu(Het(f)) = 0$.

Proof. By Lemma 3.5 we first observe that

$$Het(f) \subset \bigcup_{n\in\mathbb{N}} A(f^n).$$

Let $\mu \in \mathcal{M}(X)$ be an expansive measure of f. By Lemma 3.6, μ is also an expansive measure of f^n. Then by Lemma 3.3, we get $\mu(A(f^n)) = 0$ for all $n \in \mathbb{N}$, and so

$$\mu(Het(f)) \leq \sum_{n\in\mathbb{N}} \mu(A(f^n)) = 0.$$

This completes the proof.

□

Corollary 3.2 *Let f be a homeomorphism on a compact metric space X, and $\mu \in \mathcal{M}(X)$ be an expansive measure of f. Then $\mu(Per(f)) = 0$.*

We finish this section by describing the expansive measures of a circle homeomorphism. For this, we first show that there is no expansive measures of interval homeomorphisms.

Lemma 3.7 *A homeomorphism on a compact interval has no expansive measures.*

Proof. Let f be a homeomorphism on a compact interval I. Suppose, on the contrary, that f admits an expansive measure $\mu \in \mathcal{M}(I)$. By the continuity of f, we observe that f admits a fixed point, i.e., $Fix(f) \neq \emptyset$. Then the complement $I \setminus Fix(f)$ in I can be expressed by the disjoint union of countably many open intervals, i.e., $I \setminus Fix(f) = \bigcup_{i\in\mathbb{N}} I_i$. Since every point in I_i is a point with converging semiorbits, and so $\mu(I \setminus Fix(f)) = 0$ by Lemma 3.3. On the other hand, by Corollary 3.2, we have $\mu(\mathrm{Fix}(f)) = 0$. Hence, we get

$$\mu(I) = \mu(Fix(f)) + \mu(I \setminus Fix(f)) = 0,$$

which is a contradiction.

□

Theorem 3.3 *A circle homeomorphism has an expansive measure if and only if it is a Denjoy homeomorphism.*

Proof. Let f be a homeomorphism on the unit circle S^1, and μ be an expansive measure of f. Suppose, on the contrary, that f is not a Denjoy homeomorphism. Then either f has a periodic point or f is conjugated to a rotation (see for example [23]). We consider the following two cases.

Case 1: $Per(f) \neq \emptyset$. By Lemma 3.6, we may assume that f has a fixed point. We can cut open S^1 at the fixed point to obtain an interval I, a homeomorphism g

on I and a measure $\nu \in \mathcal{M}(I)$ corresponding to f and μ respectively. It contradicts Lemma 3.7.

Case 2: f is conjugate to a rotation. Since μ is expansive, by Remark 3.1, there is a rotation which admits an expansive measure. It contradicts Example 3.1.

Conversely, let f be a Denjoy homeomorphism on the unit circle S^1. It is well known that f has no periodic points and admits the unique minimal set Δ which is a Cantor set without isolated points [23]. By Lemma 1.2, there exists a nonatomic Borel measure ν on Δ. Moreover, by Example 2.1 in [15], we know that $f|_\Delta$ is expansive, and so ν is an expansive measure of $f|_\Delta$. We extend the measure ν on Δ to a measure μ on X by letting

$$\mu(A) = \nu(A \cap \Delta), \ \forall A \in \beta_X.$$

Then μ is an expansive measure for f.

□

3.3 Expansive Invariant Measures

In this section, we prove that every ergodic measure $\mu \in \mathcal{E}(X, f)$ with positive entropy is expansive for f. The result is based on the study in [5]. For this, we need some notations and lemmas. For given $\mu \in \mathcal{M}(X)$ and a constant $\delta > 0$, we define $\mu_\delta : X \to \mathbb{R}$ by

$$\mu_\delta(x) = \mu(\Gamma_\delta^f(x)), \ \forall x \in X.$$

For $n \in \mathbb{N}$, we define $\mu_{\delta,n} : X \to \mathbb{R}$ by

$$\mu_{\delta,n}(x) = \mu(B[x, n, \delta]), \ \forall x \in X,$$

where $B[x, n, \delta] = \{y \in X : d(f^i(x), f^i(y)) \leq \delta, \ \forall 0 \leq i \leq n-1\}$.

Lemma 3.8 *Let f be a homeomorphism on a compact metric space X, and $\mu \in \mathcal{M}(X)$. Then μ_δ is measurable for every $\delta > 0$.*

Proof. Let $\delta > 0$, $n \in \mathbb{N}$ and define

$$D_n = \{(x, y) \in X \times X : d(f^i(x), f^i(y)) \leq \delta, \quad \forall -n+1 \leq i \leq n-1\}.$$

Since f is continuous, we observe that D_n is closed in $X \times X$, and so $D_n \in \beta_{X\times X}$. Moreover, since X is compact, the product σ-algebra $\beta_X \otimes \beta_X$ on $X \times X$ satisfies $\beta_X \otimes \beta_X = \beta_{X\times X}$ (e.g., see Lemma 6.4.2 in [9]). Then $D_n \in \beta_X \otimes \beta_X$. By applying the Fubini theorem (e.g., see Theorem 3.4.1 in [9]) we obtain that the map $x \mapsto \int_X \chi_{D_n}(x, y) d\mu(y)$ is measurable, where χ_{D_n} denotes the characteristic function of

D_n. On the other hand, by the definition of D_n, for each $n \in \mathbb{N}$, we see that for each $n \in \mathbb{N}$

$$\mu_{\delta,n}(x) = \int_X \chi_{D_n}(x, y) d\mu(y),$$

and so $\mu_{\delta,n}$ is measurable. Moreover, since μ_δ is the pointwise limit of measurable functions, we obtain that it is measurable.

□

Next, we state the following basic result due to Brin and Katok [12]. Let $\mu \in \mathcal{M}(X, f)$. The *metric entropy* of μ with respect to f is defined by

$$h_\mu(f) = \sup\{h_\mu(f, P) : P \text{ is a finite measurable partition of } X\},$$

where

$$h_\mu(f, P) = -\lim_{n\to\infty} \frac{1}{n} \sum_{\xi \in P_{n-1}} \mu(\xi) \log \mu(\xi)$$

and $P_n := P \vee f^{-1}(P) \vee \cdots \vee f^{-(n-1)}(P)$ is the pullback partition of P under f^n (see [39] for details). We now recall an expression of the entropy of an ergodic measure μ, which is known as the Brin–Katok theorem.

Lemma 3.9 *Let f be a homeomorphism on a compact metric space X, and μ be a nonatomic ergodic measure of f. Then*

$$h_\mu(f) = \sup_{\delta>0} \liminf_{n\to\infty} -\frac{\log(\mu(B[x, n, \delta]))}{n}, \quad \mu\text{-a.e. } x \in X.$$

In the following theorem, we show the expansiveness of an ergodic measure with positive entropy. For this, we denote by $A \Delta B$ the symmetric difference of the sets A, B, i.e., $A \Delta B = (A \setminus B) \cup (B \setminus A)$.

Theorem 3.4 *Let f be a homeomorphism on a compact metric space X, and let $\mu \in \mathcal{E}(X, f)$. If $h_\mu(f) > 0$, then μ is expansive for f.*

Proof. Let μ be an ergodic invariant measure of f with the positive entropy $h_\mu(f) > 0$. Fix $\delta > 0$ and define

$$X_\delta = \{x \in X : \mu(\Gamma_\delta^f(x)) = 0\}.$$

Clearly, $X_\delta = \mu_\delta^{-1}(0)$. Hence, by Lemma 3.8, the set X_δ is measurable. It remains to show, by Lemma 3.1, that there exists $\delta > 0$ such that $\mu(X_\delta) = 1$.

Let $x \in X$. Since $\Gamma_\delta^f(x) = f^{-1}(\Gamma_\delta^f(f(x)))$ and μ is invariant, we have

$$\mu(\Gamma_\delta^f(x)) = \mu(\Gamma_\delta^f(f(x))).$$

It follows that $x \in X_\delta$ if and only if $f(x) \in X_\delta$. Hence, we get $f^{-1}(X_\delta) = X_\delta$.

Since $\mu(f^{-1}(X_\delta)) = \mu(X_\delta)$, it follows that X_δ is *essentially invariant*, i.e., $\mu(f^{-1}(X_\delta)\Delta X_\delta) = 0$. Since μ is ergodic, we obtain that $\mu(X_\delta) \in \{0, 1\}$ for all $\delta > 0$. Thus, it remains to show that there exists $\delta > 0$ such that $\mu(X_\delta) > 0$. To this end, we proceed as follows.

For each $\delta > 0$, define the map $\phi_\delta : X \to \mathbb{R} \cup \{\infty\}$,

$$\phi_\delta(x) = \liminf_{n\to\infty} \frac{-\log \mu(B[x, n, \delta])}{n}, \qquad \forall x \in X.$$

Set $h = \frac{h_\mu(f)}{2} > 0$, and for each $m \in \mathbb{N}$ define

$$X^m = \left\{x \in X : \phi_{\frac{1}{m}}(x) > h\right\}, \quad \forall m \in \mathbb{N}.$$

Observe that $\phi_\delta(x) \geq \phi_{\delta'}(x)$ whenever $0 < \delta < \delta'$. From this, it follows that $X^m \subset X^{m'}$ for $m \leq m'$. Moreover,

$$\{\, x \in X : \sup_{\delta>0} \phi_\delta(x) = h_\mu(f) \,\} \subset \bigcup_{m\in\mathbb{N}} X^m.$$

Then we have

$$\mu\Big(\{\, x \in X : \sup_{\delta>0} \phi_\delta(x) = h_\mu(f) \,\}\Big) \leq \lim_{m\to\infty} \mu(X^m).$$

Since μ is ergodic with positive entropy, it is a nonatomic. By Lemma 3.9, we observe that

$$\mu\Big(\{\, x \in X : \sup_{\delta>0} \phi_\delta(x) = h_\mu(f) \,\}\Big) = 1,$$

and so

$$\lim_{m\to\infty} \mu(X^m) = 1.$$

Consequently, we can take $m \in \mathbb{N}$ such that

$$\mu(X^m) > 0.$$

We will prove that $\delta = \frac{1}{m}$ works.

Let $x \in X^m$. By the definition of X^m, we obtain that $\mu(B[x, n, \delta]) < e^{-hn}$ for all sufficiently large $n \in \mathbb{N}$. Since $h > 0$, we get $\lim_{n\to\infty} \mu(B[x, n, \delta]) = 0$. Since $\mu_{\delta,n}(x) = \mu(B[x, n, \delta])$, we conclude that $\mu(\Gamma^f_\delta(x)) = 0$, and so $x \in X_\delta$. Since $x \in X^m$ is arbitrary, we obtain $X^m \subset X_\delta$ whence

$$0 < \mu(X^m) \leq \mu(X_\delta).$$

This completes the proof. □

3.4 Measure Expansive Homeomorphisms

Let f be a homeomorphism on a compact metric space X. We first observe that if f is expansive, then every nonatomic measure $\mu \in \mathcal{M}^*(X)$ is expansive for f. In this section, we study the class of homeomorphisms f for which every nonatomic measure $\mu \in \mathcal{M}^*(X)$ is expansive for f. Let us give a precise definition as follows.

Definition 3.8 We say that f is *measure expansive* if every nonatomic measure $\mu \in \mathcal{M}^*(X)$ is expansive for f.

Proposition 3.2 There are no measure expansive homeomophisms on the circle.

Proof. Suppose that there exists a circle homeomorphism f which is measure expansive. Let μ be a nonatomic measure on X. We first claim that $\text{supp}(\mu) \subset \Omega(f)$. Indeed, assume, on the contrary, that there is $x \in \text{supp}(\mu) \setminus \Omega(f)$. Since μ is expansive for f, let $\delta > 0$ be an expansive constant of μ. Since $x \notin \Omega(f)$, we may assume that the collection of open intervals $\{f^n(B(x,\delta))\}_{n\in\mathbb{Z}}$ is pairwise disjoint. Then there is $N \in \mathbb{N}$ such that the length of $f^n((B(x,\delta)))$ is less than $\frac{\delta}{2}$ for $|n| \geq N$. By the continuity of f, we can choose $\varepsilon > 0$ such that $B(x,\varepsilon) \subset \Gamma^f_\delta(x)$, and so $\mu(\Gamma^f_\delta(x)) \geq \mu(B(x,\varepsilon)) > 0$. It implies that $x \in \text{supp}(\mu)$. The contradiction shows that $\text{supp}(\mu) \subset \Omega(f)$.

By Theorem 3.3, the map f is a Denjoy homeomorphism, and hence $\Omega(f)$ is nowhere dense. Let μ denote the Lebesgue measure on the unit circle S^1. By the claim above, we have $S^1 = \text{supp}(\mu) \subset \Omega(f)$, which is a contradiction. This completes the proof.

□

We now give a characterization of measure expansive homeomorphisms on compact metric spaces. For this, we introduce the notion of countably expansive homeomorphisms.

Definition 3.9 We say that f is *countably expansive* if there is $\delta > 0$ such that the dynamic δ-ball $\Gamma^f_\delta(x)$ of f is countable for every $x \in X$.

We observe that every countably expansive homeomorphism is measure expansive. In the following theorem, due to Artigue and Carrasco-Oliveira [7], it is shown that the converse also holds.

Theorem 3.5 *Let f be a homeomorphism on a compact metric space X. Then f is measure expansive if and only if it is countably expansive.*

Proof. The sufficiency is clear, so it remains to prove the necessity. Suppose, on the contrary, that there exists a homeomorphism $f : X \to X$ on a compact metric space X which is measure expansive but not countably expansive. Then there are sequences $\delta_n \to 0$ and $x_n \in X$ such that $\Gamma^f_{\delta_n}(x_n)$ is uncountable for all $n \in \mathbb{N}$.

By Theorem 1.6, for each $n \in \mathbb{N}$ there exists a nonatomic Borel measure $\mu_n \in \mathcal{M}^*(X)$ such that $\mu_n(\Gamma^f_{\delta_n}(x_n)) = 1$. Define a Borel probability measure

$$\mu = \sum_{n \in \mathbb{N}} \frac{1}{2^n} \mu_n.$$

It is easy to see that μ is nonatomic. Since f is measure expansive, μ is expansive for f. Let $\delta > 0$ be an expansive constant of μ. Since $\delta_n \to 0$, there is $m \in \mathbb{N}$ such that $\delta_m < \delta$. Then we have

$$\mu(\Gamma^f_\delta(x_m)) \geq \mu(\Gamma^f_{\delta_m}(x_m)) \geq \frac{1}{2^m}\mu_n(\Gamma^f_{\delta_m}(x_m)) = \frac{1}{2^m} > 0.$$

The contradiction completes the proof.

□

Motivated by the notion of countable expansiveness, we introduce some level of expansiveness for homeomorphisms on compact metric spaces.

Definition 3.10 We say that f is *N-expansive* ($N \in \mathbb{N}$) if there is $\delta > 0$ such that the dynamic δ-ball $\Gamma^f_\delta(x)$ of f has at most N elements for every $x \in X$.

In the following, we construct a 2-expansive homeomorphism on a closed surface that is not expansive. The proof follows the approach of [8].

Theorem 3.6 *There is a 2-expansive homeomorphism of a closed surface which is not expansive.*

Proof. Consider the so-called derivation from the Anosov diffeomorphism $f_1 : T^2 \to T^2$ of the two-torus T^2. In particular, the nonwandering set $\Omega(f_1) = \Omega_1 \cup \{p_1\}$ is composed of a one-dimensional hyperbolic attractor Ω_1 plus a repelling fixed point p_1.

Let $f_2 = f_1^{-1}$ be the inverse of f_1 hence $\Omega(f_2) = \Omega_2 \cup \{p_2\}$, where Ω_2 is a one-dimensional hyperbolic attractor, and p_2 is an attracting fixed point.

Around p_1 and p_2 we put two disks D_1 and D_2 in the basin of repulsion and attraction of p_1 and p_2 respectively. They both are copies of the two-dimensional disk $\{x \in \mathbb{R}^2 : \|x\| \leq 2\}$ in $\mathbb{R}^2$. We choose these disks in a way that the unstable (resp. stable) foliation around p_1 (resp. p_2) corresponds to the vertical (resp. horizontal) foliation by lines in D. Since p_1 and p_2 are repelling and attracting fixed points respectively, we can assume that $f_1^{-1}|D_1$ (resp. $f_2|_{D_2}$) is the linear contraction $x \in D \mapsto \frac{1}{4}x$. Under such an assumption the annulus

$$\{x \in \mathbb{R}^2 : \frac{1}{2} \leq \|x\| \leq 2\}$$

represents a fundamental domain of f_1 (resp. f_2) in D_1 (resp. D_2).

We now remove the central disks $\{x \in D : \|x\| < \frac{1}{2}\}$ from D_1 and D_2 resulting in two annuli A_1 and A_2 (the copies of $\{x \in \mathbb{R}^2 : \frac{1}{2} \leq \|x\| \leq 2\}$ in D_1 and D_2).

In the sequel, we glue A_1 to A_2 through a suitable diffeomorphism $\psi : A_1 \to A_2$. This ψ is chosen in a way that the image of each leaf of the trace of the unstable manifold in D_1 (i.e. the vertical lines in D_1) intersects each leaf of the stable foliation in D_2 (i.e., the horizontal line in D_2) in at most two points. In particular, we obtain the following property:

(P) Two-point intersections between the image under ψ of a unstable leaf with a stable leaf are unavoidable and, moreover, such two points of intersection may be chosen to be as close as we want.

Now, the space resulting from the aforementioned gluing is a manifold, more precisely, a bi-torus bT^2. It comes equipped with a diffeomorphism $f : bT^2 \to bT^2$ equal to f_1 (resp. f_2) when restricted to the T^2-domain of f_1 and f_2 respectively. On the one hand, since the intersection points in (P) can be made arbitrarily close each other, f is not expansive. On the other hand, there is $\delta > 0$ such that $W^s_\delta(x) \cap W^u_\delta(x)$ has at most two points for every $x \in bT^2$: This is clear when $x \in \Omega_1 \cup \Omega_2$ and follows from (P) otherwise. Then, every subset $B \subset bT^2$ with $diam(f^n(B)) \leq \delta$ for all $n \in \mathbb{Z}$ has at most two points. Therefore, f is 2-expansive and the proof follows. □

In the sequel, we present an example of n-expansive homeomorphisms with shadowing on compact metric spaces, due to Carvalho and Cordeiro [14].

Theorem 3.7 *For every integer $n \geq 2$, there exists an n-expansive homeomorphism with shadowing on a compact metric space X, that is not $(n-1)$-expansive.*

Proof. Consider an expansive homeomorphism with shadowing $g : M \to M$ on a compact metric space M having infinitely many periodic points $\{p_k\}_{k\in\mathbb{N}}$. We can assume that these points belong to different orbits of g. Take any enumerable set E and define X as the disjoint union $M \cup E$. Fix a bijection $r : \mathbb{N} \to E$. Define

$$Q = \bigcup_{k\in\mathbb{N}} (\{1, \cdots, n-1\} \times \{k\} \times \{0, \cdots, \pi(p_k) - 1\}),$$

where $\pi(p_k)$ denotes the period of p_k. Notice that there is a bijection $s : Q \to \mathbb{N}$ and so the composition $q = r \circ s : Q \to E$ is a bijection too. Hence, every $x \in E$ has the form $x = q(i, k, j)$ for some $(i, k, j) \in Q$.

Denote by d_0 the metric of M and define the function $d : X \times X \to \mathbb{R}^+$ by

$$d(x, y) = \begin{cases} 0, & \text{if } x = y, \\ d_0(x, y), & \text{if } x, y \in M, \\ \frac{1}{k} + d_0(y, g^j(p_k)), & \text{if } x = q(i, k, j) \text{ and } y \in M, \\ \frac{1}{k} + d_0(x, g^j(p_k)), & \text{if } x \in M \text{ and } y = q(i, k, j), \\ \frac{1}{k}, & \text{if } x = q(i, k, j),\ y = q(l, k, j), \\ & \text{and } i \neq l, \\ \frac{1}{k} + \frac{1}{m} + d_0(g^j(p_k), g^r(p_m)), & \text{if } x = q(i, k, j),\ y = q(l, m, r), \\ & \text{and } (k, j) \neq (m, r). \end{cases}$$

We now proceed according to the following five steps.

Step 1. We claim that d is a metric of X. Indeed, we can see easily that $d(x, y) = d(y, x)$ and $d(x, y) = 0$ if and only if $x = y$. It remains to prove the triangle inequality, i.e., $d(x, z) \leq d(x, y) + d(y, z)$. It is trivial when $x, y, z \in M$. When $(x, y, z) \in M \times M \times E$ we have $z = q(i, k, j)$ and

$$d(x, z) = \frac{1}{k} + d_0(x, g^j(p_k)) \leq \frac{1}{k} + d_0(x, y) + d_0(y, g^j(p_k)) = d(x, y) + d(y, z).$$

The proof when $(x, y, z) \in M \times E \times M$ or when $(x, y, z) \in E \times M \times M$ are similar. Now suppose that $(x, y, z) \in M \times E \times E$ hence $y = q(i, k, j)$ and $z = q(l, m, r)$. If $k \neq m$ or $j \neq r$ then

$$\begin{aligned} d(x, z) &= \frac{1}{m} + d_0(x, g^r(p_m)) \\ &\leq \frac{2}{k} + \frac{1}{m} + d_0(x, g^j(p_k)) + d_0(g^j(p_k), g^r(p_m)) \\ &= d(x, y) + d(y, z). \end{aligned}$$

If $k = m$, $j = r$ and $i \neq l$ then

$$d(x, z) = \frac{1}{m} + d_0(x, g^r(p_m)) < \frac{1}{k} + \frac{1}{m} + d_0(x, g^j(p_k)) = d(x, y) + d(y, z).$$

The cases $(x, y, z) \in E \times M \times E$ or $(x, y, z) \in E \times E \times M$ can be treated similarly. Therefore, d is a metric of X.

Step 2. We prove that (X, d) is a compact metric space. For this, we show that any sequence $(x_m)_{m \in \mathbb{N}} \subset X$ has a convergent subsequence. Since (M, d_0) is compact, we can assume that $x_m \in E$ for all $m \in \mathbb{N}$. Hence $x_m = q(i_m, k_m, j_m)$ for all $m \in \mathbb{N}$. If k_m is bounded, $(x_m)_{m \in \mathbb{N}}$ is finite and so the convergent subsequence exists. Hence, we can assume that $k_m \to \infty$. Note that for all m there is $p_m \in M$ such that $d(x_m, p_m) = \frac{1}{k_m}$. Since M is compact, the sequence $(p_m)_{m \in \mathbb{N}}$ has a convergent subsequence which can be assume to be $(p_m)_{m \in \mathbb{N}}$ itself. Denote by p the corresponding limit. Thus (x_m) also converges to p. We conclude that (X, d) is compact.

We now define $f : X \to X$ by

$$f(x) = \begin{cases} g(x), & \text{if } x \in M. \\ q(i, k, (j + 1) \bmod \pi(p_k)), & \text{if } x = q(i, k, j). \end{cases}$$

It is clear that f is a homeomorphism. Moreover, E splits into an infinite number of periodic orbits $\{q(i, k, j) : 1 \leq i \leq n - 1, k \in \mathbb{N}, 1 \leq j \leq \pi(p_k) - 1\}$.

Step 3. We claim that f is n-expansive. Let δ be an expansive constant of g. Suppose that the points $x_1, \cdots, x_n, x_{n+1} \in X$ satisfy $d(f^s(x_i), f^s(x_j)) \leq \delta$ for every $s \in \mathbb{Z}$ and every $(i, j) \in \{1, \cdots, n + 1\} \times \{1, \cdots, n + 1\}$. Clearly, at most one of these points belong to M and so at least n of them belong to E. Let us prove that at least two of them are of the form $q(i,k,j)$ and $q(i,m,r)$ with $k \neq m$. Otherwise, two of them are of the form $q(i,k,j)$ and $q(i,k,r)$ with $j \neq r$. It follows that for each $s \in \mathbb{Z}$,

$$\begin{aligned} d(g^s(g^j(p_k)), g^s(g^r(p_k))) &= d(f^s(q(i,k,j)), f^s(q(i,k,r))) - \frac{2}{k} \\ &< d(f^s(q(i,k,j)), f^s(q(i,k,r))) \\ &\leq \delta, \end{aligned}$$

thus proving $g^j(p_k) = g^r(p_k)$. This would imply that $j = r$, an absurdity.

Next, we note that for every $s \in \mathbb{Z}$ it holds that

$$\begin{aligned} d(g^s(g^j(p_k)), g^s(g^r(p_m))) &= d(f^s(q(i,k,j)), f^s(q(l,m,r))) - \frac{1}{k} - \frac{1}{m} \\ &< d(f^s(q(i,k,j)), f^s(q(l,m,r))) \\ &\leq \delta. \end{aligned}$$

proving $g^j(p_k) = g^r(p_k)$ and so $p_m = p_k$, which contradicts the fact that $k \neq m$. Hence the points $\{x_1, \cdots, x_{n+1}\}$ cannot exist and so f is n-expansive. This proves the claim.

Step 4. We prove that f is not $(n-1)$-expansive. For each $\delta > 0$, choose $k \in \mathbb{N}$ such that $\frac{1}{k} < \delta$ and note that $\Gamma^f_{\frac{1}{k}}(p_k)$ contains n different points since for each $i \in \{1, \cdots, n-1\}$ the point $q(i,k,0)$ belong to it. Then f is not $(n-1)$-expansive.

Step 5. Finally, we prove that f is shadowing. Fix $\varepsilon > 0$. Since g is shadowing, there is $\delta_0 > 0$ such that every δ_0-pseudo orbit of g can be $\frac{\varepsilon}{2}$-shadowed. Choose $m \in \mathbb{N}$ such that $\frac{1}{m} < \min\{\frac{\varepsilon}{2}, \frac{\delta_0}{3}\}$ and take $\delta = \frac{1}{m}$. If $\{x_i\}_{i\in\mathbb{Z}} \subset X$ is a δ-pseudo orbit of f then either $\{x_i\}_{i\in\mathbb{Z}}$ is one of the orbits $\{q(l,k,j) : j \in \{0, \cdots, \pi(p_k)\}\}$ with $l \in \{1, \cdots, n-1\}$ and $k \in \{1, \cdots, m-1\}$ or $\{x_i\}_{i\in\mathbb{Z}}$ does not contain any point of these orbits. In the former case $\{x_i\}_{i\in\mathbb{Z}}$ can be obviously shadowed, so we will assume the latter case. Thus if $x_i = q(l,k,j)$ then $k \geq m$. Define the sequence $\{y_i\}_{i\in\mathbb{Z}} \subset M$ by y_i equal to x_i (if $x_i \in M$) or $g^j(p_k)$ (if $x_i = q(l,k,j)$).

Since

$$\begin{aligned} d_0(g(y_i), y_{i+1}) &= d(f(y_i), y_{i+1}) \\ &\leq d(f(y_i), f(x_i)) + d(f(x_i), x_{i+1}) + d(x_{i+1}, y_{i+1}) \\ &\leq \frac{1}{m} + \frac{1}{m} + \frac{1}{m} \leq \delta_g, \end{aligned}$$

we have that $\{y_i\}_{i\in\mathbb{Z}}$ is a δ_g-pseudo orbit of g. Then, there is $x \in M$ such that $d_0(g^i(x), y_i) \leq \frac{\varepsilon}{2}$ for every $i \in \mathbb{Z}$. It follows that

$$d(f^i(x), x_i) \leq d(f^i(x), y_i) + d(y_i, x_i) \leq \frac{\varepsilon}{2} + \frac{1}{m} \leq \varepsilon,$$

and so the f-orbit of x ε-shadows $\{x_i\}_{i\in\mathbb{Z}}$. This completes the proof. □

In Chap. 2, we showed that if f is expansive, then there is $\varepsilon > 0$ such that

$$W^s_\varepsilon(x, f) \subset W^s(x, f) \quad \text{and} \quad W^u_\varepsilon(x, f) \subset W^u(x, f), \quad \forall x \in X.$$

In the end of this section, we extend the result to the case of n-expansive homeomorphisms.

Let f be a homeomorphism on a compact metric space X. For $x, y \in X$, we write $x \sim y$ if $\lim_{n\to\infty} d(f^n(x), f^n(y)) = 0$. Clearly, $\sim$ is an equivalence relation on X. For given $A \subset X$, we define

$$n(A) = card\{W^s(a, f) : a \in A\}.$$

For $x \in X$ and $\varepsilon > 0$, we write $n(x, \varepsilon) = n(W^s_\varepsilon(x, f))$, and $E(x, \varepsilon)$ the collection of $n(x, \varepsilon)$ elements such that if $y, z \in E(x, \varepsilon)$ then $z \notin W^s(y, f)$. By Lemma 2.4, we observe that if f is expansive, then there is $\varepsilon > 0$ such that $n(x, \bar{\varepsilon}) = 1$ for every $x \in X$. In [14] it was posed the question whether for every n-expansive homeomorphism f on a compact metric space X, there is a constant $\varepsilon > 0$ such that

$$\sup_{x\in X} n(x, \bar{\varepsilon}) \le n, \quad \forall 0 < \bar{\varepsilon} < \varepsilon. \tag{3.4}$$

The following result by Carvalho and Cordeiro [14] gives a positive answer when f is shadowing.

Theorem 3.8 *Suppose f is an n-expansive homeomorphism with shadowing. Then there is $\varepsilon > 0$ such that for any $x \in X$, there is a sequence $\{x_1, \cdots, x_n\}$ such that*

$$W^s_\varepsilon(x, f) \subset \bigcup_{i=1}^{n} W^s(x_i, f).$$

Proof. We will show that $\varepsilon = \frac{c}{4}$ satisfies (3.4) where c be an n-expansive constant of f. Suppose, on the contrary, that it is not true, namely there are $x \in X, 0 < \bar{\varepsilon} < \varepsilon$ and a collection of distinct points $y_1, \cdots, y_{n+1} \in W^s_{\bar{\varepsilon}}(x)$ such that $y_i \notin W^s(y_j, f)$ for all distinct $i, j \in \{1, \cdots, n+1\}$. This implies that there are $0 < r < \bar{\varepsilon}$ and a sequence $\{a^{(i,j)}_m\}_{m\in\mathbb{N}}$ for distinct $i, j \in \{1, \cdots, n+1\}$ such that $a^{(i,j)}_m = a^{(j,i)}_m$, $a^{(i,j)}_m \to \infty$ as $m \to \infty$ and

$$d(f^{a^{(i,j)}_m}(y_i), f^{a^{(i,j)}_m}(y_j)) > r, \quad \forall m \in \mathbb{N}.$$

We will get a contradiction through the following cases.

Case 1: There are $r' > 0$ and a sequence of positive integers $\{a_m\}_{m\in\mathbb{N}}$ such that $a_m \to \infty$ as $m \to \infty$ such that

$$d(f^{a_m}(y_i), f^{a_m}(y_j)) > r', \quad \forall m \in \mathbb{N}, i \ne j \in \{1, \cdots, n+1\}. \tag{3.5}$$

By compactness we can assume without loss of generality that $f^{a_m}(x) \to x_0$ and $f^{a_m}(y_i) \to x_i$ for some $x_0, x_1, \cdots, x_{n+1} \in X$ as $m \to \infty$. For each $i \in \{1, \cdots, n+1\}$ and each $l \in \mathbb{Z}$ since $y_i \in W^s_{\bar{\varepsilon}}(x)$ one has

$$d(f^l(x_i), f^l(x_0)) = \lim_{m\to\infty} d(f^l(f^{a_m}(y_i)), f^l(f^{a_m}(x)) \leq \bar{\varepsilon} < c.$$

Since c is an n-expansive constant, $x_i = x_j$ for some distinct $i, j \in \{1, \cdots, n+1\}$. However, by letting $m \to \infty$ in (3.5) we get

$$d(x_i, x_j) = \lim_{m\to\infty} d(f^{a_m}(y_i), f^{a_m}(y_j)) \geq r' > 0,$$

which is a contradiction. This finishes the proof in Case 1.

Case 2: The situation in Case 1 does not occur. Then, there are distinct $i, j \in \{1, \cdots, n+1\}$ and a positive integer sequence $b_m \to \infty$ (as $m \to \infty$) such that

$$d(f^{b_m}(y_i), f^{b_m}(y_j)) < \frac{1}{m}, \quad \forall m \in \mathbb{N}.$$

We can assume

$$b_m < a_m^{(i,j)} < b_{m+1}, \quad \forall m \in \mathbb{N}.$$

Take $\alpha = \min\{\bar{\varepsilon}, \frac{r}{4}\}$ and choose $\beta > 0$ corresponding to α by the shadowing of f. Choose $w \in \mathbb{N}$ such that

$$\frac{1}{w} < \beta.$$

Given $l \in \{1, \cdots, n+1\}$ we define the sequence $\{x_k^l\}_{k\in\mathbb{Z}}$ by

$$x_k^l = \begin{cases} f^k(y_i), \text{ if } k < b_{w+l} \text{ or } k \geq b_{w+l+1} \\ f^k(y_j), \text{ if } b_{w+l} \leq k < b_{w+l+1}. \end{cases}$$

These sequences are β-pseudo orbits. Indeed, for $k = b_{w+l} - 1$ we have

$$\begin{aligned} d(f(x^l_{b_{w+l-1}}), x^l_{b_{w+l}}) &= d(f(f^{b_{w+l}-1}(y_i), f^{b_{w+l}}(y_j))) \\ &= d(f^{b_{w+l}}(y_i), f^{b_{w+l}}(y_j)) \\ &< \frac{1}{w+l} < \frac{1}{w} < \beta, \end{aligned}$$

and for $k = b_{w+l+1} - 1$ we have

$$\begin{aligned} d(f(x^l_{b_{w+l+1}-1}), x^l_{b_{w+l+1}}) &= d(f(f^{b_{w+l+1}-1}(y_j), f^{b_{w+l+1}}(y_i)) \\ &= d(f^{b_{w+l+1}}(y_j), f^{b_{w+l+1}}(y_i)) \\ &< \frac{1}{w+l+1} < \frac{1}{w} < \beta. \end{aligned}$$

Then, the shadowing provides $z_1, \cdots, z_{n+1} \in X$ such that

$$d(f^k(z_l), x_k^l) \leq \alpha, \quad \forall l = \{1, \cdots, n+1\}, \forall k \in \mathbb{Z}.$$

By taking $p = a_{w+t}^{(i,j)}$ and distinct $t, s \in \{1, \cdots, n+1\}$ one has

$$\begin{aligned} d(f^p(z_t), f^p(z_s)) &\geq d(x_p^t, x_p^s) - d(x_p^t, f^p(z_t)) - d(f^p(z_s), x_p^s) \\ &= d(f^p(y_j), f^p(y_i)) - d(f^p(z_t), x_p^t) - d(f^p(z_s), x_p^s) \\ &\geq r - 2\alpha \geq \frac{r}{2} > 0. \end{aligned}$$

Then, the points $z_1, \cdots, z_{n+1}$ are distinct. Also, since for $k \leq 0$ one has

$$\begin{aligned} d(f^k(z_1), f^k(z_l)) &\leq d(f^k(z_1), f^k(y_i)) + d(f^k(y_i), f^k(z_l)) \\ &\leq d(f^k(z_l), x_k^1) + d(x_k^l, f^k(z_l)) \\ &\leq 2\alpha \leq 2\bar{\varepsilon} \leq c \end{aligned}$$

and for $k > 0$ one has

$$d(x_k^1, x_k^l) \leq d(x_k^1, x) + d(x, x_k^l) \leq 2\bar{\varepsilon},$$

hence

$$\begin{aligned} d(f^k(z_1), f^k(z_l)) &\leq d(f^k(z_1), x_k^1) + d(x_k^1, x_k^l) + d(x_k^l, f^k(z_l)) \\ &\leq \alpha + 2\bar{\varepsilon} + \alpha \leq 4\bar{\varepsilon} \leq c. \end{aligned}$$

Then we conclude that $d(f^k(z_1), f^k(z_l)) \leq c$ for all $k \in \mathbb{Z}$ and $\{1, \cdots, n+1\}$ contradicting that c is an n-expansive constant of f. This finishes the proof. □

From this result we obtain the following consequence.

Corollary 3.3 *Let $f : X \to X$ be an n-expansive homeomorphism with shadowing on a compact metric space. Then, for every $x \in X$ there is $\varepsilon > 0$ such that $W_\varepsilon^s(x, f) \subset W^s(x, f)$ and $W_\varepsilon^u(x, f) \subset W^u(x, f)$.*

Proof. Let $\varepsilon > 0$ be given in Theorem 3.8 such that $n(x, \varepsilon) \leq n$ for every $x \in X$. It follows from the definition that there is $E(x, \varepsilon) \subset W_\varepsilon^s(x, f)$ with $n(x, \varepsilon)$ points such that any two distinct points in $E(x, \varepsilon)$ belong to different stable sets. Moreover, every $y \in W_\varepsilon^s(x, f)$ belong to the stable set of some $z \in E(x, \varepsilon)$. We can assume that $x \in E(x, \varepsilon)$. Thus, if $z \in E(x, \varepsilon) \setminus \{x\}$, there is $r_z > 0$ and a sequence of positive integers $\{a_m^z\}_{m \in \mathbb{N}}$ such that $a_m^z \to \infty$ as $m \to \infty$ and

$$d(f^{a_m^z}(z) f^{a_m^z}(x)) > r_z, \quad \forall m \in \mathbb{N}.$$

Take

$$\varepsilon_x = \frac{1}{4} \min\{r_z : z \in E(x, \varepsilon) \setminus \{x\}\}.$$

Then, if $y \in W^s_{\varepsilon_x}(x)$, $z \in E(x, \varepsilon) \setminus \{x\}$ and $m \in \mathbb{N}$ one has

$$d(f^{a^z_m}(z), f^{a^z_m}(y)) \geq d(f^{a^z_m}(z), f^{a^z_m}(x)) - d(f^{a^z_m}(x), f^{a^z_m}(y)) \geq r_z - \varepsilon_x \geq \frac{r_z}{2},$$

proving $y \notin W^s(z, f)$, so $y \in W^s(x, f)$. Consequently, $W^s_{\varepsilon_x}(x, f) \subset W^s(x, f)$.

Similarly, we can show that $W^u_\varepsilon(x, f) \subset W^u(x, f)$, and so the proof is complete. □

Chapter 4
Shadowing and Topological Stability

In this chapter, we study the shadowing and topologically stable measures for a homeomorphism f on a compact metric space X. In Sect. 4.1, we introduce the notion of a shadowing measure for f, which is applied to characterize the shadowing homeomorphisms. Moreover, we show that if a Borel measure $\mu \in \mathcal{M}(X)$ with maximal entropy is shadowing for f and there is a nonminimal point $x \in \text{supp}(\mu)$ then $h_\mu(f) > 0$. In Sect. 4.2, we introduce a notion of topologically stable measures for homeomorphisms, and present a measurable version of Walters' stability theorem which says that any expansive measure with shadowing is topologically stable. Finally, we study the topologically stable measures for circle homeomorphisms.

4.1 Shadowing Measures

In this section, we introduce the notions of shadowing measures and shadowable points of a homeomorphism on a compact metric space, and use them to characterize the notion of shadowing for homeomorphisms. Throughout this section, let f be a homeomorphism on a compact metric space X. A δ-pseudo orbit $\xi = \{x_i\}_{i \in \mathbb{Z}}$ is said to be through a subset B of X if $x_0 \in B$. In particular, we say that ξ is through $x \in X$ if $B = \{x\}$.

Definition 4.1 A Borel measure $\mu \in \mathcal{M}(X)$ is *shadowing* for f (or f is μ-shadowing) if for every $\varepsilon > 0$ there are $\delta > 0$ and a Borel set $B \subset X$ with $\mu(X \setminus B) = 0$ such that every δ-pseudo orbit $\xi = \{x_n\}_{n \in \mathbb{Z}}$ through B is ε-shadowed by a point in X.

Definition 4.2 We say that f is *measure shadowing* (resp. *invariant measure shadowing*) on an invariant subset A of X if μ is shadowing for $f|_A$ for all $\mu \in \mathcal{M}(A)$ (resp. $\mu \in \mathcal{M}(A, f|_A)$).

The following proposition characterizes the shadowing measures using their supports. For this, we first introduce an auxiliary concept as follows.

K. Lee et al., *Topological Dynamics from a Measure-Theoretic Viewpoint*, Springer Asia Pacific Mathematics Series 14, https://doi.org/10.1007/978-981-95-6390-6_4

Definition 4.3 We say that f is *shadowing through a subset B of X* if for every $\varepsilon > 0$, there is $\delta > 0$ such that every δ-pseudo orbit of f through B is ε-shadowed by a point in X.

Proposition 4.1 *Let f be a homeomorphism on a compact metric space X. A Borel measure $\mu \in \mathcal{M}(X)$ is shadowing for f if and only if f is shadowing through $\mathrm{supp}(\mu)$.*

Proof. If f is shadowing through $\mathrm{supp}(\mu)$, μ is shadowing for f by taking $B = \mathrm{supp}(\mu)$ in the corresponding definition. Conversely, suppose that μ is shadowing for f. For $\varepsilon > 0$, take a constant $0 < \delta < \varepsilon/2$ and a Borel subset B of X such that $\mu(B) = 1$ and any 2δ-pseudo orbit of f through B can be $\frac{\varepsilon}{2}$-shadowed. Since $\mu(B) = 1$, $\mathrm{supp}(\mu) \subset \overline{B}$.

Now let $\xi = \{x_n\}_{n\in\mathbb{Z}}$ be a δ-pseudo orbit through $\mathrm{supp}(\mu)$. Since $\mathrm{supp}(\mu) \subset \overline{B}$, we can take $y \in B$ such that

$$d(y, x_0) < \delta \quad \text{and} \quad d(f(y), f(x_0)) < \delta.$$

Then the sequence $\{y_n\}_{n\in\mathbb{Z}}$ defined by

$$y_n = \begin{cases} y, & \text{if } n = 0, \\ x_n, & \text{otherwise,} \end{cases}$$

is a 2δ-pseudo-orbit of f through B. Hence, there is $z \in X$ such that

$$d(f^n(z), y_n) < \frac{\varepsilon}{2}, \ \forall n \in \mathbb{Z}.$$

Since

$$d(z, x_0) \le d(z, y) + d(y, x_0) < \varepsilon,$$

we derive that ξ is ε-shadowed by $z \in X$. Hence f is shadowing through $\mathrm{supp}(\mu)$. □

In the following example, we construct a homeomorphism f on a compact metric space X such that f is not shadowing but it admits a shadowing measure of f.

Example 4.1 Let $X = \{\infty\} \cup (\mathbb{Z} \times \{0\}) \cup (\bigcup_{n\in\mathbb{N}}\{-n, \cdots, n\} \times \{\frac{1}{n}\})$ be a subspace of the sphere $\mathbb{R}^2 \cup \{\infty\}$ and consider a homeomorphism f on X defined by

$$f(x) = \begin{cases} \infty, & \text{if } x = \infty, \\ (n+1, 0), & \text{if } x = (n, 0), \\ (i+1, \frac{1}{2}), & \text{if } x = (i, \frac{1}{2}), \\ (-n, \frac{1}{2}), & \text{if } x = (n, \frac{1}{2}). \end{cases}$$

where $-n \leq i < n$.

Suppose that f is shadowing. Let $\varepsilon < \frac{1}{4}d((0,0), \infty)$ and take $\delta > 0$ such that every δ-pseudo orbit of f is ε-shadowed by a point in X. Take $n \in \mathbb{N}$ such that $d((-n, 0), (n, 0)) < \delta$. Let $\{x_i\}_{i\in\mathbb{Z}}$ be a sequence in X given by $x_i = f^i((0, 0))$ for $i \leq n-1$, $x_{n+2kn+i} = f^i((-n, 0))$ for all $0 \leq i \leq 2n-1$ and $k \in \mathbb{N}$. Since f is shadowing, there is $y \in X$ such that

$$d(f^i(y), x_i) < \varepsilon, \ \forall i \in \mathbb{Z}.$$

Since $x_i \to \infty$ as $i \to -\infty$, we see that y is not a periodic point of f. It implies that $f^i(y) \to \infty$ as $i \to \infty$, and so, we have

$$\begin{aligned}\varepsilon &> d(f^{n+2kn+n}(y), (0, 0)) \\ &\geq d((0, 0), \infty) - d(\infty, f^{n+2kn+n}(y)) \to d((0, 0), \infty) \text{ as } k \to \infty.\end{aligned}$$

The contradiction shows that f is not shadowing.

Let μ be a Borel measure on X. We claim that μ is shadowing for f if and only if $\operatorname{supp}(\mu) \cap (\{\infty\} \cup (\mathbb{Z} \times \{0\})) = \emptyset$. In fact, let μ be a shadowing measure for f and suppose that there is $x \in \operatorname{supp}(\mu) \cap (\{\infty\} \cup (\mathbb{Z} \times \{0\}))$. Let $\varepsilon < \frac{1}{4}d((0, 0), \infty)$. By Proposition 4.1, we observe that there is $\delta > 0$ such that any δ-pseudo orbit through $\operatorname{supp}(\mu)$ can be ε-shadowed by a point in X. Consider a δ-pseudo orbit through x given by $x_i \to \infty$ as $i \to -\infty$, and there is $n \in \mathbb{N}$ such that $x_{n+2kn+i} = x_i$ for all $0 \leq i \leq 2n-1$ and $k \in \mathbb{N}$. Then we observe that there is no point $y \in X$ such that

$$d(f^i(y), x_i) < \varepsilon, \ \forall i \in \mathbb{Z}.$$

The contradiction shows that $\operatorname{supp}(\mu) \cap (\{\infty\} \cup (\mathbb{Z} \times \{0\})) = \emptyset$.

Conversely, suppose that $\operatorname{supp}(\mu) \cap (\{\infty\} \cup (\mathbb{Z} \times \{0\})) = \emptyset$. There is $m \in \mathbb{N}$ such that

$$\operatorname{supp}(\mu) \subset \bigcup_{n=1}^{m} (\{-n, \ldots, n\} \times \{\tfrac{1}{n}\}).$$

Let $\varepsilon > 0$. We take $\delta > 0$ such that

$$B(x, \delta) = \{x\}, \qquad \forall x \in \bigcup_{n=1}^{m} (\{-n, \ldots, n\} \times \{\tfrac{1}{n}\}).$$

Let $\xi = \{x_i\}_{i\in\mathbb{Z}}$ be a δ-pseudo orbit of f through $\operatorname{supp}(\mu)$. By the choice of δ, we see that ξ is a real orbit of f, and so it is ε-shadowed by x_0. Hence μ is shadowing for f.

We remark that if $\mu \in \mathcal{M}(X)$ is shadowing for f, then any point $x \in \operatorname{supp}(\mu)$ satisfies the following property: for every $\varepsilon > 0$, there is $\delta > 0$ such that every δ-pseudo orbit $\{x_i\}_{i\in\mathbb{Z}}$ through x is ε-shadowed by a point in X. We call this property the shadowableness of a point. Precisely, we state the definition of a shadowable point of a homeomorphism as follows.

Definition 4.4 A point $x \in X$ is *shadowable* for f if for every $\varepsilon > 0$ there is $\delta > 0$ such that every δ-pseudo orbit $\{x_i\}_{i\in\mathbb{Z}}$ of f through x is ε-shadowed by a point in X. We denote by $Sh(f)$ the set of shadowable points of f.

Remark 4.1 The set of shadowable points is a Borel set (see [24]). We observe that a point $x \in X$ is shadowable for f if and only if the Dirac measure δ_x centered at x is shadowing for f.

Clearly if f is shadowing, then every point is shadowable for f. We observe that the converse is true on compact metric spaces by the following proposition.

Proposition 4.2 *Let f be a homeomorphism on a compact metric space X. Then f is shadowing if and only if $Sh(f) = X$.*

Proof. Suppose that $Sh(f) = X$. We claim that f is shadowing. Suppose not. Then there are a constant $\varepsilon > 0$ and a sequence $\{\xi^k\}_{k\in\mathbb{N}}$ of $\frac{1}{k}$-pseudo orbits $\xi^k = \{\xi^k_n\}_{n\in\mathbb{Z}}$ such that ξ^k is not 2ε-shadowed for every $k \in \mathbb{N}$.

Since X is compact, we may assume that ξ^k_0 converges to p for some $p \in X$. Since p is shadowable, we can take $\delta > 0$ such that every δ-pseudo orbit through p is ε-shadowed by a point in X. For each $k \in \mathbb{N}$, we define the sequence $\hat{\xi}^k = \{\hat{\xi}^k_n\}_{n\in\mathbb{Z}}$ by

$$\hat{\xi}^k_n = \begin{cases} \xi^k_n, & \text{if } n \neq 0, \\ p, & \text{if } n = 0. \end{cases}$$

It is clear that $\hat{\xi}^k$ is through p for all $k \in \mathbb{N}$. Since

$$d\big(f(\hat{\xi}^k_{n-1}), \hat{\xi}^k_n\big) = \begin{cases} d\big(f(\xi^k_{n-1}), \xi^k_n\big), & \text{if } n \neq 0, 1, \\ d\big(f(p), \xi^k_1\big), & \text{if } n = 1, \\ d\big(f(\xi^k_{-1}), p\big), & \text{if } n = 0, \end{cases}$$

we have

$$d\big(f(\hat{\xi}^k_{n-1}), \hat{\xi}^k_n\big) \leq \begin{cases} \dfrac{1}{k}, & \text{if } n \neq 0, 1, \\ d\big(f(p), f(\xi^k_0)\big) + \dfrac{1}{k}, & \text{if } n = 1, \\ d\big(\xi^k_0, p\big) + \dfrac{1}{k}, & \text{if } n = 0. \end{cases}$$

Since f is continuous, there is a constant $k_0 > 0$ such that $\frac{1}{k} < \frac{\delta}{2}$, $d(\xi^k_0, p) < \frac{\delta}{2}$ and $d(f(p), f(\xi^k_0)) < \frac{\delta}{2}$ for all $k \geq k_0$. Then we see that $\{\hat{\xi}^k_n\}_{n\in\mathbb{Z}}$ is a δ-pseudo orbit of f for all $k \geq k_0$. Since p is shadowable, for each $k \geq k_0$, there exists $x_k \in X$ such that $d(f^n(x_k), \hat{\xi}_n^k) \le \varepsilon$ for all $n \in \mathbb{Z}$. It follows that $d(f^n(x_k), \xi_n^k) \le \varepsilon$ for $n \neq 0$. Moreover, we have

$$d(x_k, \xi_0^k) \le d(x_k, p) + d(p, \xi_0^k) \le \varepsilon + d(p, \xi_0^k) < 2\varepsilon, \qquad \forall k \ge k_0.$$

Consequently, we obtain that ξ^k is 2ε-shadowed by x_k for all $k \ge k_0$. The contradiction shows that f is shadowing. This completes the proof.

□

Finally, we show that f is shadowing if and only if it is μ-shadowing for every $\mu \in \mathcal{M}(X, f)$ (or $\mu \in \mathcal{M}(X)$). Let $c > 0$. We denote by $Sh_c(f)$ the collection of $x \in X$ such that there is a constant $\delta = \delta(x) > 0$ such that any δ-pseudo orbit through x is c-shadowed by a point in X.

Lemma 4.1 *Let f be a homeomorphism on a compact metric space X, and constants $0 < c < b$. If $x \in Sh_c(f)$, then there exists $\delta = \delta(x) > 0$ such that every δ-pseudo orbit $\{x_i\}_{i \in \mathbb{Z}}$ of f with $d(x, x_0) < \delta$ is b-shadowed by a point in X.*

Proof. Let $x \in Sh_c(f)$. Then there exists $\delta_1 > 0$ such that every δ_1-pseudo orbit $\{z_i\}_{i \in \mathbb{Z}}$ through x is c-shadowed by a point in X. By the continuity of f, we take a constant $0 < \delta < \min\{\frac{\delta_1}{2}, b - c\}$ such that if $d(y, z) < \delta$ for $y, z \in X$, then $d(f(y), f(z)) < \frac{\delta_1}{2}$.

Let $\{x_i\}_{i \in \mathbb{Z}}$ be a δ-pseudo orbit of f with $d(x, x_0) < \delta$. We define a sequence $\{y_i\}_{i \in \mathbb{Z}}$ by $y_0 = x$ and $y_i = x_i$ for $i \neq 0$. Since

$$\begin{aligned} d(f(y_0), y_1) = d(f(x), x_1) &\le d(f(x), f(x_0)) + d(f(x_0), x_1) \\ &\le \frac{\delta_1}{2} + \delta < \delta_1, \end{aligned}$$

and

$$\begin{aligned} d(f(y_{-1}), y_0) = d(f(x_{-1}), x) &\le d(f(x_{-1}), x_0) + d(x_0, x) \\ &\le \frac{\delta_1}{2} + \delta < \delta_1, \end{aligned}$$

we observe that $\{y_i\}_{i \in \mathbb{Z}}$ is a δ-pseudo orbit of f through x. Since $x \in Sh_c(f)$, there is $p \in X$ such that $\{y_i\}_{i \in \mathbb{Z}}$ is c-shadowed by p. Then we get

$$d(f^i(p), x_i) = d(f^i(p), y_i) < c < b, \ \forall i \neq 0,$$

and

$$d(p, x_0) \le d(p, x) + d(x, x_0) < b.$$

Hence $\{x_i\}_{i \in \mathbb{Z}}$ is b-shadowed by $p \in X$. This completes the proof.

□

Lemma 4.2 *Let f be a homeomorphism on a compact metric space X, $x \in X$, and a constant $c > 0$. If $\overline{\mathcal{O}_f(x)} \cap Sh_c(f) \neq \emptyset$, then $x \in Sh_b(f)$ for every $b > c$. In particular, if $\overline{\mathcal{O}_f(x)} \cap Sh(f) \neq \emptyset$, then $x \in Sh(f)$.*

Proof. Let $y \in \overline{\mathcal{O}_f(x)} \cap Sh_c(f)$, and a constant $b > c$. By Lemma 4.1, there exists a constant $\delta > 0$ such that every δ-pseudo orbit $\{y_i\}_{i\in\mathbb{Z}}$ of f with $d(y_0, y) < \delta$ is b-shadowed by a point in X. Since $y \in \overline{\mathcal{O}_f(x)}$, we can take $N \in \mathbb{N}$ such that $d(f^N(x), y) < \frac{\delta}{2}$. By the continuity of f, for given $N \in \mathbb{N}$, there is $0 < \delta' < \delta$ such that for every δ'-pseudo orbit $\{y_i\}_{i\in\mathbb{Z}}$ of f with $y_0 = x$, $d(f^N(x), y_N) < \frac{\delta}{2}$.

Let $\{x_i\}_{i\in\mathbb{Z}}$ be a δ'-pseudo orbit of f with $x_0 = x$. Since

$$d(y, x_N) \leq d(y, f^N(x)) + d(f^N(x), x_N) < \delta,$$

by Lemma 4.1, we obtain that $\{x_i\}_{i\in\mathbb{Z}}$ is b-shadowed by a point in X, and so $x \in Sh_b(f)$.

□

In the following theorem, we show a characterization of shadowing for homeomorphisms by using shadowing measures.

Theorem 4.1 *Let f be a homeomorphism on a compact metric space X. Then the following are pairwise equivalent:*

(1) f is shadowing;
(2) f is measure shadowing;
(3) f is invariant measure shadowing;
(4) $\mu(Sh(f)) = 1$ for all $\mu \in \mathcal{M}(X, f)$.

Proof. (1) $\Rightarrow$ (2) $\Rightarrow$ (3) $\Rightarrow$ (4). It is clear from the definitions.

(4) $\Rightarrow$ (1). Suppose that $\mu(Sh(f)) = 1$ for all $\mu \in \mathcal{M}(X, f)$. Let $x \in X$. Since $\omega(x, f)$ is a nonempty closed invariant set, by Theorem 1.7, there exists $\mu \in \mathcal{M}(X, f)$ such that $\mu(\omega(x, f)) = 1$. Since $\mu(Sh(f)) = 1$, we have $\omega(x, f) \cap Sh(f) \neq \emptyset$. By Lemma 4.2, we get $x \in Sh(f)$, and so $Sh(f) = X$. By Proposition 4.2, we derive that f is shadowing.

□

In the rest of this section, we show that if a Borel measure $\mu \in \mathcal{M}(X)$ is shadowing for f with maximal entropy and there is a nonminimal point $x \in \text{supp}(\mu)$ then $h_\mu(f) > 0$. The result is based on [25]. For this, we need some notations. Given $n \in \mathbb{N}, \delta > 0$, we say that $E \subseteq X$ is (n, δ)-separated if $B_n^+[x, \delta] \cap E = \{x\}, \forall x \in E$ where

$$B_n^+[x, \delta] = \bigcap_{i=0}^{n-1} f^{-i}(B[f^i(x), \delta]).$$

It follows from the compactness of X that the number

$$S(n, \delta) = \max\{\text{Card}(E) : E \text{is}(n, \delta)\text{-separeated}\}$$

is finite. Define the topological entropy of f by the Bowen–Dinaburg formula

$$h_{top}(f) = \lim_{\delta \to 0} \lim_{n \to \infty} \sup \frac{1}{n} \log S(n, \delta).$$

Here we recall the relationship between topological entropy and metric entropy, which is known as variational principle, as follows:

$$h_{top}(f) = \sup\{h_\mu(f) : \mu \in \mathcal{M}(X, f)\}.$$

A measure $\mu \in \mathcal{M}(X, f)$ is called a measure of maximal entropy for f if $h_{top}(f) = h_\mu(f)$.

With these definitions we can state the following theorem.

Theorem 4.2 *Let f be a homeomorphism on a compact metric space X, and $\mu \in \mathcal{M}(X)$ be a Borel measure of maximal entropy for f. If μ is shadowing for f and there is a nonminimal point $x \in \text{supp}(\mu)$, then $h_\mu(f) > 0$.*

The proof uses the following notation and facts. We say that a δ-chain $\{x_i\}_{i=0}^R$ is a δ-cycle through x if $x_0 = x_R = x$. For given $e > 0$, a pair of δ-cycles $(\{z_i^0\}_{i=0}^m, \{z_i^1\}_{i=0}^m)$ through x is said to be e-separated if $d(z_i^0, z_i^1) > e$ for some $0 < i < m$. The number m is called the period of the cycle.

Remark 4.2 If $\{z_i^0\}_{i=0}^R$ and $\{z_i^1\}_{i=0}^l$ are δ-cycles through z with $d(z_j^0, z_j^1) > \varepsilon$ for some $0 < j < \min\{R, l\}$ then the pair

$$\{z_0^0, z_1^0, \ldots, z_j^0, \ldots, z_{R-1}^0, z_0^1, z_1^1, \ldots, z_{l-1}^1, z_0^1\}, \text{ and}$$

$$\{z_0^1, z_1^1, \ldots, z_j^1, \ldots, z_{l-1}^1, z_0^0, z_1^0, \ldots, z_{R-1}^0, z_0^0\}$$

is an e-separated pair of δ-cycles of period $R + l$.

With these notations we have the following lemma.

Lemma 4.3 *Let f be a homeomorphism on a compact metric space X. Given $e > 0$ and $z \in X$, if there are $x \in X$ and a closed invariant set S such that $z \in S \subseteq \omega(x, f)$ and $\omega(x, f) \setminus B(S, e) \neq \emptyset$, then for every $\delta > 0$, there exists an e-separated pair of δ-cycles through z.*

Proof. Fix $p \in \omega(x, f)$ with $d(p, S) > e$. Given $\delta > 0$, since $f|_{\omega(x,f)}$ is chain transitive, there is a δ-chain $\{x_i^1\}_{i=0}^a$ of $f|_{\omega(x,f)}$ such that $x_0^1 = z$ and $x_a^1 = p$. Since $z \in S$ and S is invariant, $f^a(z) \in S$, then

$$d(f^a(z), p) \geq d(p, S) > e.$$

Again by chain transitivity of $f|_{\omega(x,f)}$, there is a pair of δ-chains $(\{y_i^0\}_{i=0}^b, \{y_i^1\}_{i=0}^c)$ such that

$$(y_0^0, y_0^1) = (f^a(z), p) \quad (y_b^0, y_c^1) = (z, z).$$

Now consider the δ-cycles

$$\{z, f(z), \ldots, f^{a-1}(z), y_0^0, y_1^0, \ldots, y_{b-1}^0, z\}$$

and

$$\{z, x_1^1, \ldots, x_{a-1}^1, y_0^1, y_1^1, \ldots, y_{c-1}^b, z\}.$$

Since $d(y_0^0, y_0^1) = d(f^a(z), p) > e$ we have that these δ-cycles form an e-separated pair of δ-cycles. This complete the proof.

□

Next, we obtain positive topological entropy using pairs of δ-cycles.

Lemma 4.4 *Let $b > 0$, $\delta > 0$, $e > 2b$ and $x \in X$. Suppose that every δ-pseudo orbit of f through x is b-shadowed by a point in X. If there exists an e-separated pair of δ-cycles through $z \in \omega(x, f)$, then $h_{top}(f) > 0$.*

Proof. By hypothesis there exist a pair of δ-cycles $(\{z_i^0\}_{i=0}^m, \{z_i^1\}_{i=0}^m)$ which is e-separated of period $m \in \mathbb{N}^+$. Fix $0 < j < n$ such that

$$d(z_j^0, z_j^1) > e. \tag{4.1}$$

Since $z \in \omega(x, f)$, we can take $k > 0$ such that $d(f^k(x), z) \leq \delta$. Given $n \in \mathbb{N}$ and $s = (s_1, \ldots, s_n) \in \{0, 1\}^n$ we can consider the following δ-chain:

$$\{x, f(x), \ldots, f^{k-1}(x), z_0^{s_1}, z_1^{s_1}, \ldots, z_{n-1}^{s_1}, \ldots, z_0^{s_n}, z_1^{s_n}, \ldots z_{n-1}^{s_n}\}.$$

Since every δ-pseudo orbit through x can be b-shadowed, there is a map

$$y : \{0, 1\}^n \to B_b(x),$$

so that $y(s)$ is a point which b-shadows the above δ-chain for all $s \in \{0, 1\}^n$.

Note that this map is injective. Otherwise, for two different elements of $\{0, 1\}^n$ the corresponding δ-chains can be b-shadowed by the same y. Then, we would have $d(z_j^0, p) < b$ and $d(z_j^1, p) < b$ for some p in the orbit of y yielding

$$d(z_j^0, z_j^1) \leq d(z_j^0, p) + d(z_j^1, p) < 2b < \varepsilon,$$

contradicting (4.1).

Let $E_n = y(\{0, 1\}^n)$. Since y is injective, we have

$$\text{Card}(E_n) = \text{Card}(\{0, 1\}^n) = 2^n.$$

We claim that E_n is $(k+mn, e-2b)$-separated. Indeed, for every $s, t \in \{0,1\}^n$, if $s \neq t$, then $s_a \neq t_a$ for some $1 \leq a \leq n$. Letting $K = k + (a-1)m + j$, we have $0 \leq K \leq k+mn$ (because $1 \leq a \leq n$ and $0 \leq j \leq m$) and also

$$\begin{aligned} d(f^K(y(s)), f^K(y(t))) &\geq d(z_j^{s_a}, z_j^{s_a}) - d(f^K(y(s)), z_j^{s_a}) - d(f^K(y(t)), z_j^{t_a}) \\ &> e - 2b. \end{aligned}$$

Now noting that $\mathrm{Card}(E_n) = 2^n$ we get $s(k+mn, e-2b) \geq 2^n$, $\forall n \in \mathbb{N}$. Then

$$\begin{aligned} h_{\mathrm{top}}(f) &\geq \limsup_{n\to\infty} \frac{1}{n} \log s(k+mn, \varepsilon - 2\beta) \\ &\geq \limsup_{n\to\infty} \frac{1}{k+mn} \log s(k+mn, \varepsilon - 2\beta) \\ &\geq \limsup_{n\to\infty} \frac{n \log 2}{k+mn} \\ &= \frac{1}{m} \log 2 \ > \ 0. \end{aligned}$$

□

Proof of Theorem 4.2. Suppose that μ is shadowing with maximal entropy, and let $x \in \mathrm{supp}(\mu)$ be a point whose ω-limit set $\omega(x, f)$ is nonminimal. Since $\omega(x, f)$ is nonminimal, there exists a compact invariant set $S \subset \omega(x, f)$ and $\varepsilon > 0$ such that $\omega(x, f) \backslash B_e(S) \neq \emptyset$. Take any $z \in S$. Then the assumptions of Lemma 4.3 are satisfied. Fix $0 < b < \frac{e}{2}$, and choose $\delta > 0$ corresponding to the shadowableness of x for b.

Therefore, x is a (δ, b)-shadowable point. By Lemma 4.3, there exists an e-separated pair of δ-cycles through z. Hence, by Lemma 4.4, we have $h_{top}(f) > 0$, and consequently $h_\mu(f) > 0$.

□

4.2 Topologically Stable Measures

In this section, we show a measurable version of Walters' stability theorem. For this, we first introduce the notion of topological stability of Borel measures for a homeomorphism on a compact metric space. The contents of this section are based on [27]. Throughout this section, let f be a homeomorphism on a compact metric space X, and $\mu \in \mathcal{M}(X)$. For $A \subset X$ and $\varepsilon > 0$, we define

$$B(A, \varepsilon) = \bigcup_{x \in A} B(x, \varepsilon).$$

We denote by 2^X the space of all compact nonempty subsets of X with the Hausdorff metric d_H, i.e.,

$$d_H(A, B) = \inf\{\varepsilon > 0 : A \subset B(B, \varepsilon) \text{ and } B \subset B(A, \varepsilon)\}.$$

A map $H : \mathrm{Dom}(H) \to 2^X$ is referred to as a compact-valued map of X. We write $d(H, Id) \le \varepsilon$ for some $\varepsilon > 0$ if $H(x) \subset B[x, \varepsilon]$, where $B[z, \varepsilon]$ denotes the closed ε-ball centered at $z \in X$.

Definition 4.5 Let $H : \mathrm{Dom}(H) \to 2^X$ be a compact-valued map.

(1) A map H is said to be *upper semicontinuous* if for every $x \in \mathrm{Dom}(H)$ and any $\varepsilon > 0$, there is $\delta > 0$ such that if $d(x, y) < \delta$, then $H(y) \subset B(H(x), \varepsilon)$.
(2) A map H is said to be *lower semicontinuous* if for every $x \in \mathrm{Dom}(H)$ and any $\varepsilon > 0$, there is $\delta > 0$ such that if $d(x, y) < \delta$, then $H(x) \subset B(H(y), \varepsilon)$.
(3) A map H is said to be *continuous* if it is both upper semicontinuous and lower semicontinuous.

Definition 4.6 A Borel measure $\mu \in \mathcal{M}(X)$ is said to be *topologically stable* for f if for every $\varepsilon > 0$ there is $\delta > 0$ such that for every homeomorphism g on X with $d_{C^0}(f, g) \le \delta$, there is an upper semicontinuous map $H : \mathrm{Dom}(H) \to 2^X$ with g-invariant domain such that:

(1) $\mu(X \setminus \mathrm{Dom}(H)) = 0$,
(2) $\mu(H(x)) = 0$ for all $x \in \mathrm{Dom}(H)$,
(3) $d(H, Id) \le \varepsilon$,
(4) $H(g(x)) = f(H(x))$ for all $x \in \mathrm{Dom}(H)$.

It is known that for an expansive homeomorphism f, every nonatomic measure is expansive for f. In the following example, we will show a similar result for topological stability. More precisely, if a homeomorphism f is topologically stable, then every nonatomic measure $\mu \in \mathcal{M}^*(X)$ is topologically stable for f. We note that the result fails for atomic measures as in Theorem 4.4(1).

Example 4.2 Let f be a topologically stable homeomorphism on a compact metric space X (see Definition 2.10). Then every nonatomic Borel measure $\mu \in \mathcal{M}^*(X)$ is topologically stable for f. In fact, let $\mu \in \mathcal{M}^*(X)$ and $\varepsilon > 0$. We take $\delta > 0$ corresponding to ε by the topological stability of f. Let g be a homeomorphism on X with $d_{C^0}(f, g) \le \delta$. Then there is a continuous map $h : X \to X$ such that $d_{C^0}(h, Id_X) \le \varepsilon$ and $f \circ h = h \circ g$. We define a set-valued map $H : X \to 2^X$ by $H(x) = \{h(x)\}$ for all $x \in X$. It is clear that $\mathrm{Dom}(H) = X$ and H is a compact-valued map. Moreover, since h is continuous, we have H is upper semicontinuous.

On the other hand, it is clear that $\mu(X \setminus \mathrm{Dom}(H)) = 0$. Since μ is nonatomic, we have $\mu(H(x)) = \mu(\{h(x)\}) = 0$ for all $x \in X$. Moreover, we have $d(H(x), x) = d(h(x), x) \le \varepsilon$ for all $x \in X$. Since $f \circ h = h \circ g$, we derive that $f \circ H = H \circ g$. Therefore μ is topologically stable for f.

In Chap. 2, we proved Walters' stability theorem, which says that every expansive homeomorphism with shadowing on a compact metric space is topologically stable (see Theorem 2.7). In the following theorem, we present a measurable version of Walters' stability theorem.

Theorem 4.3 *Let f be a homeomorphism on a compact metric space X and $\mu \in \mathcal{M}(X)$. If μ is expansive and shadowing for f, then it is topologically stable for f.*

Proof. Suppose $\mu \in \mathcal{M}(X)$ is expansive, shadowing measure for f. Let $e > 0$ be an expansive constant of μ, $\varepsilon > 0$ and $0 < \varepsilon' < \min\{\frac{e}{2}, \varepsilon\}$. By the shadowing of μ, we take a constant $\delta > 0$ corresponding to $\varepsilon' > 0$. Let g be a homeomorphism on X with $d_{C^0}(f, g) \leq \delta$. For each $x \in X$, we define

$$H(x) = \{y \in X : d(f^n(y), g^n(x)) \leq \varepsilon', \ \forall n \in \mathbb{Z}\}, \ \forall x \in X.$$

and denote by $\mathrm{Dom}(H)$ the collection of $x \in X$ such that $H(x)$ is nonempty. We note that $\mathrm{Dom}(H)$ is nonempty. In fact, let $x \in B$. Since $d_{C^0}(f, g) \leq \delta$, the g-orbit $\{g^n(x)\}_{n\in\mathbb{Z}}$ is a δ-pseudo orbit of f. By the shadowing of μ, there is $y \in X$ such that

$$d(f^n(y), g^n(x)) \leq \varepsilon', \ \forall n \in \mathbb{Z}.$$

Then $H(x) \neq \emptyset$ for all $x \in B$, and so $\mathrm{Dom}(H)$ is nonempty.

Moreover, $H(x)$ is compact for every $x \in \mathrm{Dom}(H)$. Indeed, let $\{y_i\}_{i\in\mathbb{N}}$ be a sequence in $H(x)$ such that $y_i \to y$ for some $y \in X$. Since

$$d(f^n(y), g^n(x)) = \lim_{i\to\infty} d(f^n(y_i), g^n(x)) \leq \varepsilon', \ \forall n \in \mathbb{Z},$$

we get $y \in H(x)$, and so $H(x)$ is compact. Hence we can define a compact-valued map $H : \mathrm{Dom}(H) \to 2^X$.

We first show that $\mathrm{Dom}(H)$ is g-invariant. In fact, let $x \in \mathrm{Dom}(H)$ and $y \in H(x)$. Since

$$d(f^n(f(y)), g^n(g(x))) \leq \varepsilon', \ \forall n \in \mathbb{Z},$$

we get $f(y) \in H(g(x))$, and so $g(x) \in \mathrm{Dom}(H)$.

Next, we show that H is upper semicontinuous. Suppose not. Then there are $x \in \mathrm{Dom}(H)$ and a constant $\varepsilon > 0$ such that for each $n \in \mathbb{N}$, there is $y_n \in \mathrm{Dom}(H)$ with $d(x, y_n) < \frac{1}{n}$ and $H(y_n) \setminus B(H(x), \varepsilon) \neq \emptyset$, where $B(H(x), \varepsilon) = \bigcup_{z\in H(x)} B(z, \varepsilon)$. Let $z_n \in H(y_n) \setminus B(H(x), \varepsilon)$. Since X is compact, we may assume that $z_n \to z \in X$ as $n \to \infty$. It is clear that $z \notin H(x)$.

For each $i \in \mathbb{Z}$, we have

$$\begin{aligned} d(f^i(z), g^i(x)) &= \lim_{n\to\infty} d(f^i(z_n), g^i(x)) \\ &\leq \lim_{n\to\infty} (d(f^i(z_n), g^i(y_n)) + d(g^i(y_n), g^i(x))) \\ &\leq \varepsilon'. \end{aligned}$$

Then $z \in H(x)$. The contradiction shows that H is upper semicontinuous.

(1) $\mathrm{Dom}(H)$ is measurable, and $\mu(X \setminus \mathrm{Dom}(H)) = 0$. We will prove that $\mathrm{Dom}(H)$ is closed in X. In fact, let $\{x_k\}_{k\in\mathbb{N}}$ be a sequence in $\mathrm{Dom}(H)$ which converges to $x \in X$. Since $x_k \in \mathrm{Dom}(H)$, we can choose a sequence $\{y_k \in H(x_k)\}_{k\in\mathbb{N}}$ such that

$$d(f^n(y_k), g^n(x_k)) \le \varepsilon', \quad \forall k \in \mathbb{N} \text{ and } n \in \mathbb{Z}. \tag{4.2}$$

Since X is compact, we may assume that $y_k \to y$ as $k \to \infty$ for some $y \in X$. For fixed $n \in \mathbb{Z}$, by letting $k \to \infty$ in (4.2), we get $d(f^n(y), g^n(x)) \le \varepsilon'$. It implies that $y \in H(x)$, and so $x \in \mathrm{Dom}(H)$. Hence $\mathrm{Dom}(H)$ is closed in X and so measurable. Moreover, we have $\mu(X \setminus \mathrm{Dom}(H)) \le \mu(X \setminus B) = 0$.

(2) $\mu(H(x)) = 0$ for all $x \in X$. It is clear that $\mu(H(x)) = 0$ if $x \notin \mathrm{Dom}(H)$. Let $x \in \mathrm{Dom}(H)$ and $y \in H(x)$. For $z \in H(x)$, we have

$$\begin{aligned} d(f^n(z), f^n(y)) &\le d(f^n(z), f^n(x)) + d(f^n(x), f^n(y)) \\ &\le 2\varepsilon' < e, \ \forall n \in \mathbb{Z}. \end{aligned}$$

Then we have $z \in \Gamma_e^f(y)$, and so $H(x) \subset \Gamma_e^f(y)$. Hence, we get $\mu(H(x)) \subset \mu(\Gamma_e^f(y)) = 0$ since μ is expansive for f.

(3) $d(H, Id) \le \varepsilon$. It is clear by the definition of H.

(4) $H(g(x)) = f(H(x))$ for all $x \in \mathrm{Dom}(H)$. Let $x \in \mathrm{Dom}(H)$. We have

$$\begin{aligned} f(H(x)) &= f\left(\bigcap_{n\in\mathbb{Z}} f^{-n}\big(B[g^n(x), \varepsilon']\big)\right) \\ &= \bigcap_{n\in\mathbb{Z}} f^{-n+1}\big(B[g^n(x), \varepsilon']\big) \\ &= \bigcap_{n\in\mathbb{Z}} f^{-n}\big(B[g^{n+1}(x), \varepsilon']\big) \\ &= \bigcap_{n\in\mathbb{Z}} f^{-n}\big(B[g^n(g(x)), \varepsilon']\big) \\ &= H(g(x)). \end{aligned}$$

Then $H(g(x)) = f(H(x))$ for all $x \in \mathrm{Dom}(H)$. This completes the proof. □

Theorem 4.3 implies the abundance of homeomorphisms for which every expansive measure is topologically stable. We denote by $\mathrm{Homeo}(X)$ the space of homeomorphisms on X equipped with the topology induced by the metric d_{C^0}. Note that $\mathrm{Homeo}(X)$ is a Baire space.

Definition 4.7 A subset R of a topological space X is said to be *residual* if it contains an intersection of countably many open and dense subsets in X. We say that generically, a homeomorphism on X satisfies a property P if there is a residual subset R of Homeo(X) such that every $f \in R$ satisfies the property P.

Corollary 4.1 *Generically, a homeomorphism f on a compact manifold without boundary satisfies that every expansive measure is topologically stable for f.*

Proof. It was proved in [34] that generically every homeomorphism $f \in$ Homeo(M) on a compact manifold without boundary is shadowing, that is, there is residual subset $\mathcal{R}$ of Homeo(M) such that every $f \in \mathcal{R}$ is shadowing. Let $f \in \mathcal{R}$ and $\mu \in \mathcal{M}(M)$ be an expansive measure of f. Since f is shadowing, we see that μ is shadowing, and so it is topologically stable by Theorem 4.3.

□

We observe that the result in Corollary 4.1 is false without the genericity hypothesis. It means there is a homeomorphism on a certain closed manifolds exhibiting an expansive measure which is not topologically stable. See Example 4.3 below.

In the rest of this section we study topologically stable measures for circle homeomorphisms. Let f be a homeomorphism on the unit circle S^1, and denote by Leb the Lebesgue measure on S^1. We say that p is an isolated periodic point of f if $p \in Per(f)$ is isolated from $Per(f)$. An isolated fixed point p is said to be *topologically hyperbolic* for f if p is a sink or source of f. An isolated periodic point p is said to be *topologically hyperbolic* for f if p is a topologically hyperbolic fixed point of f^{2k}, where k denotes the period of p.

We recall that every Denjoy homeomorphism f on the circle is uniquely ergodic, i.e., f has a unique invariant measure μ (see, for example, Theorem 11.2.9 [23]). Moreover, supp(μ) is the unique minimal set of f, and the complement of supp(μ) is called the *gaps* of f. Recall that a Borel measure μ is said to be supported on A if supp$(\mu) \subset A$.

With all the notations in mind, we state the main result of this section as follows.

Theorem 4.4 *Let $f : S^1 \to S^1$ be a homeomorphism.*

(1) If f is topologically stable, then the Dirac measure δ_p is topologically stable for f if and only if $p \notin \Omega(f)$.
(2) If the Lebesgue measure Leb is topologically stable for f, then $Per(f) \neq \emptyset$.
(3) If $\Omega(f) = S^1$, then f has no topologically stable measures.
(4) If f is a Denjoy homeomorphism and $\mu \in \mathcal{M}(S^1)$ is a topologically stable measure of f, then μ is supported on the gaps of f. In particular, the unique invariant measure of f is expansive but it is not topologically stable.
(5) If $\mu \in \mathcal{M}(S^1)$ is topologically stable for f and p is an isolated periodic point of f in $supp(\mu)$, then p is topologically hyperbolic.
(6) Suppose $Per(f)$ is finite. Then the Lebesgue measure Leb is topologically stable for f if and only if f is topologically stable.

Define a set valued map $\varphi :$ Homeo$(X) \to 2^X$ by $\varphi(f) = CR(f)$ for all $f \in$ Homeo(X). In the following lemma, we show that φ is a continuous map.

Lemma 4.5 *The map φ is continuous. In particular, if $p \notin CR(f)$ for some $f \in$ Homeo(X), there is $\delta = \delta(f, p) > 0$ such that $p \notin CR(g)$ for every $g \in$ Homeo(X) with $d_{C^0}(f, g) < \delta$.*

Proof. Suppose that φ is not upper semicontinuous at $f \in$ Homeo(X). Then there exist a constant $\varepsilon > 0$ and a sequence $\{g_n\}_{n\in\mathbb{N}}$ in Homeo(X) such that $d_{C^0}(g_n, f) \to 0$ as $k \to \infty$, and for each $n \in \mathbb{N}$, there is $x \in \varphi(g_n)$ with $d(x_n, \varphi(f)) \geq \varepsilon$. Since X is compact, we may assume that $x_n \to x_0$ for some $x_0 \in X$. It is clear that $x_0 \notin \varphi(f) = CR(f)$.

Let $\delta > 0$. We take $k \in \mathbb{N}$ such that $d_{C^0}(g_k, f) < \frac{\delta}{3}$, $d(x_k, x_0) < \frac{\delta}{3}$ and $d(f(x_k), f(x_0)) < \frac{\delta}{3}$. Since $x_k \in \varphi(g_k)$, there is a $\delta/3$-chain $\{y_0 = x_k, y_1, \cdots, y_m = x_k\}$ of g_k from x_k to itself. Consider the following sequence $z_0 = z_m = x_0$, and $z_i = y_i$ for $1 \leq i \leq m - 1$. Since

$$d(f(z_0), z_1) \leq d(f(x_0), f(x_k)) + d(f(x_k), g_k(x_k)) + d(g_k(x_k), y_1) \leq \delta$$

and

$$d(f(z_{m-1}), z_m) \leq d(f(x_{m-1}), g_k(x_{m-1})) + d(g_k(x_{m-1}), x_k) + d(x_k, x_0) \leq \delta,$$

we see that $\{z_i\}_{i=0}^m$ is a δ-chain from x_0 to itself. Since $\delta > 0$ is arbitrary, we get $x_0 \in CR(f)$. The contradiction shows that φ is upper semicontinuous on Homeo(X).

On the other hand, we suppose that φ is not lower semicontinuous at $f \in$ Homeo(X). Then there exist a constant $\varepsilon > 0$ and a sequence $\{g_n\}_{n\in\mathbb{N}}$ in Homeo(X) such that $d_{C_0}(g_n, f) \to 0$ as $n \to \infty$, and for each $n \in \mathbb{N}$ there is $x \in \varphi(f) = CR(f)$ with $d(x_n, CR(g_n)) \geq \varepsilon$. Since $CR(f)$ is compact, we may assume that $x_n \to x_0$ for some $x_0 \in CR(f)$.

Let $\delta > 0$. We take $k \in \mathbb{N}$ such that $d_{C^0}(g_k, f) < \delta/3$, $d(x_k, x_0) < \delta/3$ and $d(f(x_k), f(x_0)) < \delta/3$. Since $x_0 \in CR(f)$, there is a $\delta/3$-chain $\{y_0 = x_0, y_1, \cdots, y_m = x_0\}$ of f from x_0 to itself. Consider the following sequence $z_0 = x_k = z_m$ and $z_i = y_i$ for all $1 \leq i \leq m - 1$. Since

$$d(g_k(z_0), z_1) \leq d(g_k(z_0), f(z_0)) + d(f(z_0), f(x_0)) + d(f(x_0), y_1) \leq \delta,$$

and

$$d(g_k(z_{m-1}), z_m) \leq d(g_k(z_{m-1}), f(z_{m-1})) + d(f(z_{m-1}), x_0) + d(x_0, x_k) \leq \delta,$$

we see that $\{z_i\}_{i=0}^m$ is a δ-chain of g_k from x_k to itself. Since $\delta > 0$ is arbitrary, we get $x_k \in CR(g_k)$. The contradiction shows that φ is lower semicontinuous on Homeo(X).

Consequently, we derive that φ is continuous.

□

Lemma 4.6 *Let f be a homeomorphism on the unit circle S^1, and let $p \in S^1 \setminus CR(f)$. If the Dirac measure δ_p is shadowing for f, then δ_p is topologically stable for f.*

Proof. Let $p \in S^1 \setminus CR(f)$ be such that the Dirac measure δ_p is shadowing for f. Since $p \notin \Omega$, there is a neighborhood U of p such that $f^m(U) \cap f^n(U)$ for every $m \neq n \in \mathbb{Z}$. Then there exists a constant $c > 0$ such that

$$f^n(z) \notin [p-c, p+c] \text{ for all } z \in [p-c, p+c] \text{ and } n \in \mathbb{Z} \setminus \{0\}. \tag{4.3}$$

Let $0 < \varepsilon < 1$. Since f is a circle homeomorphism, we can take $c > 0$ smaller such that if $z \in [p-c, p+c]$ then

$$d(f^n(z), f^n(p)) \leq \frac{\varepsilon}{2}, \forall n \in \mathbb{Z}. \tag{4.4}$$

Since $p \notin CR(f)$, by Lemma 4.5, there is $\delta > 0$ such that $p \notin CR(g)$ for every $g \in \text{Homeo}(S^1)$ that is δ-close to f. Let g be a homeomorphism on S^1 such that $d_{C^0}(f, g) < \delta$. Then we have

$$g^n(p) \neq g^m(p), \ \forall n \neq m \in \mathbb{Z}. \tag{4.5}$$

Moreover, since δ_p is shadowing for f, p is shadowable for f. Then we can take $\delta > 0$ smaller such that every δ-pseudo orbit through p is $\frac{\min\{\varepsilon, c\}}{2}$-shadowed by a point in X. We claim that there exists a point $z \neq p$ in $[p-c, p+c]$ such that

$$d(f^n(z), g^n(p)) \leq \varepsilon \text{ for all } n \in \mathbb{Z}. \tag{4.6}$$

Indeed, it is clear that the orbit $\{g^n(p)\}_{n \in \mathbb{Z}}$ is a δ-pseudo orbit of f through p. Since p is shadowable, there is $y \in S^1$ such that

$$d(f^n(y), g^n(p)) \leq \frac{\min\{\varepsilon, c\}}{2} \text{ for all } n \in \mathbb{Z}. \tag{4.7}$$

If $y \neq p$, we can choose $z = y$, and then (4.6) follows from (4.7). If $y = p$, by (4.7), we have $d(f^n(p), g^n(p)) \leq \frac{\varepsilon}{2}$ for all $n \in \mathbb{Z}$. We take $z \in [p-c, p+c]$ such that $z \neq p$. By (4.4) we have

$$d(f^m(z), g^n(p)) \leq \varepsilon \text{ for all } n \in \mathbb{Z}.$$

This completes the proof of the claim.

Next, we define a set valued map H on S^1 by

$$H(x) = \begin{cases} \emptyset, \text{ if } x \notin g\text{-orbit of } p \\ \{f^n(\xi)\}, \text{ if } x = g^n(p) \text{ for some } n \in \mathbb{Z}. \end{cases}$$

By (4.5), we see that H is well defined. Let $\text{Dom}(H) = \mathcal{O}_g(p)$. It is clear that $H : \text{Dom}(H) \to 2^X$ is a compact-valued map, and $\text{Dom}(H)$ is g-invariant. Moreover, by the definition of H, we observe that $p \in \text{Dom}(H)$, and $\delta_p(S^1 \setminus \text{Dom}(H)) = 0$.

Next, we prove that $\delta_p \circ H = 0$, i.e., $p \notin H(x)$ for all $x \in S^1$. Suppose, on the contrary, that $p \in H(x)$ for some $x \in S^1$. By the definition of H, we get $x = g^n(p) \in \mathcal{O}_g(p)$ for some $n \in \mathbb{Z}$, and so $p = f^n(z)$. Since $z \in [p - c, p + c]$, by (4.3), we obtain that $x = p = z$, which contradicts (4.6). Then $\delta_p \circ H = 0$.

We prove that H is upper semicontinuous. Let $x \in \mathrm{Dom}(H)$ and U be a neighborhood of $H(x)$. Then there is $n \in \mathbb{Z}$ such that $x = g^n(p)$ and $H(x) = \{f^n(z)\} \subset U$. Let $\{x_k\}_{k\in\mathbb{N}}$ be a sequence such that $x_k \to x$ as $k \to \infty$. We may assume that $x_k \in \mathrm{Dom}(H)$ for all $k \in \mathbb{N}$. For each $k \in \mathbb{N}$, there is $n_k \in \mathbb{Z}$ such that $x_k = g^{n_k}(p)$ and $H(x_k) = \{f^{n_k}(z)\}$. Since $g^{n_k - n}(p) = g^{-n}(x_k) \to g^{-n}(x) = p$ as $k \to \infty$, and $p \notin CR(g)$, there is $k_0 \in \mathbb{N}$ such that $n_k = n$ for all $k \geq k_0$. Then $H(x_k) = \{f^{n_k}(z)\} = \{f^n(z)\} \subset U$ for all $k \geq k_0$. Hence H is upper semicontinuous.

Next, we prove $f \circ H = H \circ g$. Let $x \in \mathcal{O}_g(p)$, i.e., $x = g^n(p)$ for some $n \in \mathbb{Z}$. Then we have

$$f(H(x)) = f(\{f^n(z)\}) = \{f^{n+1}(z)\} = H(g^{n+1}(p)) = H(g(x)).$$

Hence we get $f \circ H = H \circ g$.

Finally, if $x = g^n(p)$ for some $n \in \mathbb{Z}$, we have $H(x) = \{f^n(z)\}$. By (4.6), we have $H(x) \subset B[g^n(p), \varepsilon]$. It follows that $d(H, Id) \leq \varepsilon$. This completes the proof.

□

Corollary 4.2 *Let f be a homeomorphism on the unit circle S^1, and let $p \in S^1 \setminus \Omega(f)$. If f is shadowing, then the Dirac measure δ_p is topologically stable for f.*

Proof. Since f is shadowing, by Theorem 2.6 we have $\Omega(f) = CR(f)$. Moreover, by Theorem 4.1, every Dirac measure is shadowing for f. Then the result follows from Lemma 4.6.

□

Since every topologically stable homeomorphism of S^1 is shadowing (see, for example, Remark 2.4.12 [3]), we obtain the following corollary.

Corollary 4.3 *If a circle homeomorphism f is topologically stable, then the Dirac measure δ_p is topologically stable for every $p \notin \Omega(f)$.*

It is known that every topologically stable homeomorphism on a closed manifold admits a periodic point (see, for example, Theorem 2.4.8 [3]). Therefore, there does not exist a topologically stable minimal homeomorphism on closed manifolds. It is natural to consider the existence of topologically stable measures of a minimal homeomorphism. In the following lemma, we give a negative approach to the problem. For this, we recall that f is *periodic* if $Per(f) = X$.

Lemma 4.7 *A minimal homeomorphism on an infinite compact metric space that can be approximated by periodic homeomorphisms admits no topologically stable measures.*

Proof. Let f be a minimal homeomorphism on a compact metric space X such that f is approximated by periodic homeomorphisms. Suppose, on the contrary, that there is a topologically stable measure $\mu \in \mathcal{M}(X)$ of f. Let δ correspond to $\varepsilon = 1$ by the topological stability of μ.

We first claim that if $g \in \text{Homeo}(X)$ with $d_{C^0}(f, g) \leq \delta$ then g is aperiodic with respect to μ, i.e., $\mu(Per(g)) = 0$. Suppose not. Then there is $g \in \text{Homeo}(X)$ such that $d_{C^0}(f, g) \leq \delta$ and $\mu(Per(g)) > 0$. Since μ is topologically stable, there exists an upper semicontinuous compact-valued map $H : \text{Dom}(H) \to 2^X$ with g-invariant domain such that $\mu(X \setminus \text{Dom}(H)) = 0$, $\mu(H(x)) = 0$ for all $x \in \text{Dom}(H)$, and $f \circ H = H \circ g$. Since $\mu(Per(g)) > 0$ and $\mu(X \setminus \text{Dom}(H)) = 0$, we have $Per(f) \cap \text{Dom}(H) \neq \emptyset$. Let $x \in Per(g) \cap \text{Dom}(H)$, and n be the period of x. We define

$$A = \bigcup_{i=0}^{n-1} H(g^i(x)).$$

We observe that A is compact and f-invariant. In fact,

$$\begin{aligned} f(A) &= f\left(\bigcup_{i=0}^{n-1} H(g^i(x))\right) = \bigcup_{i=0}^{n-1} f\big(H(g^i(x))\big) \\ &= \bigcup_{i=0}^{n-1} H(g^{i+1}(x)) = \bigcup_{i=0}^{n-1} H(g^i(x)) = A. \end{aligned}$$

Since $\mu(H(x)) = 0$ for all $x \in Dom(H)$, we have

$$\mu(A) \leq \sum_{i=0}^{n-1} \mu(H(g^i(x))) = 0. \tag{4.8}$$

Moreover, since $x \in \text{Dom}(H)$, there is $y \in H(x) \subset A$. Since f is minimal, we see that $\mathcal{O}_f(y)$ is dense in X, and so $A \subset X = \overline{\mathcal{O}_f(y)} \subset A$. By (4.8) we have $\mu(X) = \mu(A) = 0$. The contradiction completes the proof of the claim.

By hypothesis, there is a periodic homeomorphism $g \in \text{Homeo}(X)$ such that $d_{C^0}(f, g) \leq \delta$. Since g is periodic, we have $Per(g) = X$. By the claim above, we have $\mu(X) = \mu(Per(g)) = 0$. The contradiction completes the proof.

□

Lemma 4.8 *Let f be a homeomorphism on a compact metric space X. If a Borel measure $\mu \in \mathcal{M}(X)$ is topologically stable for f, then for any $\varepsilon > 0$ there exists $\delta > 0$ such that for every homeomorphism g on X with $d_{C^0}(f, g) < \delta$, the g-orbit of x can be ε-shadowed under f for μ-almost everywhere $x \in X$.*

Proof. Let $\varepsilon > 0$, and take $\delta > 0$ corresponding to ε by the topological stability of μ. Let g be a homeomorphism on X with $d_{C^0}(f, g) \leq \delta$. By the topological stability of μ, there is an upper semicontinuous map $H : \mathrm{Dom}(H) \to 2^X$ with g-invariant domain such that

(1) $\mu(X \setminus \mathrm{Dom}(H)) = 0$,
(2) $\mu(H(x)) = 0$ for $x \in \mathrm{Dom}(H)$,
(3) $d(H, Id) \leq \varepsilon$,
(4) $H(g(x)) = f(H(x))$ for $x \in \mathrm{Dom}(H)$.

Then for each $x \in \mathrm{Dom}(H) \subset X$, there is $y \in H(x)$ such that

$$f^n(y) \subset f^n(H(x)) = H(g^n(x)) \subset B[g^n(x), \varepsilon], \ \forall n \in \mathbb{Z}.$$

Then we get $d(f^n(y), g^n(x)) \leq \varepsilon$ for all $n \in \mathbb{Z}$. This completes the proof.

□

Lemma 4.9 *Every rotation map on the unit circle S^1 has no topologically stable measures.*

Proof. Let R_q be a rotation map of angle q on the unit circle S^1. We consider the following two cases: q is rational or q is irrational. If q is irrational, then R_q is minimal and its is approximated by periodic homeomorphisms R_p where R_p is a rational rotation. By Lemma 4.7, we obtain the result.

In the rest of proof, we suppose that q is rational. From this we can choose $\varepsilon > 0$ such that for each $y \in S^1$, the union $\bigcup_{m\in\mathbb{Z}} R_q^m([y-\varepsilon, y+\varepsilon])$ is a proper compact invariant subset of S^1. We claim that if g is an irrational rotation, then there is no g-orbit which can be ε-shadowed under R_q. Assume, on the contrary, that there exist $x, y \in S^1$ such that

$$d(R_q^n(y), g^n(x)) \leq \varepsilon.$$

Then $g^n(y) \in [R_q^n(y) - \varepsilon, R_q^n(y) + \varepsilon]$ for every $n \in \mathbb{Z}$. Moreover, since R_q is an isometry, we have

$$\begin{aligned} [R_q^n(y) - \varepsilon, \ R_q^n(y) + \varepsilon] &= R_q^n([\, y - \varepsilon, \ y + \varepsilon \,]) \\ &\subset \bigcup_{m\in\mathbb{Z}} R_q^m([\, y - \varepsilon, \ y + \varepsilon \,]), \qquad \forall n \in \mathbb{Z}. \end{aligned}$$

Therefore,

$$g^n(x) \in \bigcup_{m\in\mathbb{Z}} R_q^m([y - \varepsilon, y + \varepsilon]), \quad \forall n \in \mathbb{Z}.$$

Since $\bigcup_{m\in\mathbb{Z}} R_q^m([y-\varepsilon, y+\varepsilon])$ is a proper compact subset of S^1, the limit set $\omega(x, g)$ is a proper subset of S^1. On the other hand, since g is minimal, we get $\omega(x, g) = S^1$. The contradiction completes the proof of the claim.

Suppose, on the contrary, that R_q has a topologically stable measure $\mu \in \mathcal{M}(X)$. By Lemma 4.8 we choose $\delta > 0$ corresponding to ε. Take an irrational number p close to q such that $g = R_p$ is the irrational rotation satisfying $d(R_q, g) \leq \delta$.

Then by Lemma 4.8 there exists a Borel set $B \subset S^1$ such that $\mu(S^1 \setminus B) = 0$ and every g-orbit of $x \in B$ can be ε-shadowed under R_q. Moreover, by the claim above, there is no g-orbit can be ε-shadowed under R_q. From this we get $B = \emptyset$, and so $\mu(S^1) = \mu(S^1 \setminus B) = 0$. The contradiction completes the proof. □

Our next result is precisely item (4) of Theorem 4.4.

Lemma 4.10 *Every topologically stable measure of a Denjoy homeomorphism f is supported on the gaps of f.*

Proof. Let f be a Denjoy homeomorphism on the unit circle S^1 and $\mu \in \mathcal{M}(S^1)$. Suppose, on the contrary, that μ is not supported on the gaps of f, that is, $supp(\mu) \cap M_f \neq \emptyset$ where M_f denotes the minimal set of f. Let $\varepsilon > 0$ be such that $B[x, \varepsilon]$ is an interval for every $x \in S^1$. Take $\delta > 0$ corresponding to ε by the topological stability of μ.

We claim that every homeomorphism $g \in \text{Homeo}(S^1)$ δ-close to f is aperiodic with respect to μ. Suppose not. Then there is a homeomorphism g on S^1 with $d_{C^0}(f, g) \leq \delta$ such that g is not aperiodic with respect to μ, i.e., $\mu(Per(g)) > 0$. Since μ is topologically stable for f, there is an upper semicontinuous compact-valued map $H : \text{Dom}(H) \to 2^{S^1}$ with g-invariant domain such that $\mu(S^1 \setminus \text{Dom}(H)) = 0$, $\mu(H(x)) = 0$ for all $x \in \text{Dom}(H)$, $d_{C^0}(H, Id) \leq \varepsilon$ and $f \circ H = H \circ g$. Then we see that $Per(g) \cap \text{Dom}(H) \neq \emptyset$. Let $x \in Per(g) \cap \text{Dom}(H)$ and $n \in \mathbb{N}$ be the period of x. Note that $H(x)$ is a nonempty compact f^n-invariant set which is contained in $B[x, \varepsilon]$. Since $B[x, \varepsilon]$ is an interval, we can take $a = \min H(x)$ and $b = \max H(x)$. Since $H(x)$ is invariant for f^n which in turn is orientation preserving, we obtain that a and b are periodic points of f. Note that every Denjoy homeomorphism has no periodic points. Then we get a contradiction, and so it completes the proof of the claim.

Since $\text{supp}(\mu) \cap M_f \neq \emptyset$, we can choose $x \in \text{supp}(\mu) \cap M_f$. Now take an open I interval such that $x \in I$ and length of I is less than $\varepsilon/2$. By the continuity of f, we take a constant $\varepsilon_1 > 0$ such that $d(y, z) < \varepsilon_1$ for $y, z \in S^1$ implies that $d(f(y), f(z)) < \varepsilon/2$. Take $n \in \mathbb{N}$ as the smallest number such that $f^n(I) \cap I \neq \emptyset$. Choose $y \in Y$ such that $f^n(y) \in I$. Next, we can take a homeomorphism R_1 of S^1 such that $R_1(z) = z$ if $z \notin I$, $d_{C^0}(R_1, id) < \varepsilon_1$ and $R_1(x) = y$ (see for instance Lemma 0.3.1 in [33]). Similarly, we take a homeomorphism R_2 of S^1 such that $R_2(z) = z$ if $z \notin I$, $d_{C^0}(R_2, id) < \frac{\varepsilon}{4}$ and $R_2(f^n(y)) = x$.

We define a homeomorphism on S^1 by $g_0 = R_2 \circ f \circ R_1$. We observe that g_0 satisfies that $g_0^i(x) = f^i(x)$ for $0 \leq i \leq n-1$ and $g_0^n(x) = R_2(f^n(y)) = x$. It means $x \in Per(g_0)$. Moreover, we note that

$$d_{C^0}(f, g_0) \leq d_{C^0}(f, f \circ R_1) + d_{C^0}(f \circ R_1, R_2 \circ f \circ R_1) < \frac{\varepsilon}{2}.$$

By another perturbation, we can find another homeomorphism $g : S^1 \to S^1$ with $d(g_0, g) \leq \frac{\delta}{2}$ such that there exists an open interval I such that $x \in I \subset Per(g)$.

Since $x \in \text{supp}(\mu)$, we have $\mu(I) > 0$. Consequently, there is a homeomorphism g with $d(f, g) \leq d(f, g_0) + d(g_0, g) \leq \delta$ satisfying $\mu(Per(g)) \geq \mu(I) > 0$. Since this contradicts the claim, we obtain the result.

□

The next example is the proof of the second part of item (4) of Theorem 4.4.

Example 4.3 By Corollary 4.1, we know that generically for a homeomorphism f on a closed manifold, every expansive measure is topologically stable for f. We observe that the fact is false without the genericity hypothesis. Indeed, every Denjoy homeomorphism f has the unique ergodic measure μ which is expansive for f by Theorem 3.3. Note that μ is supported on the minimal set. Then μ is not topologically stable by Lemma 4.10.

The next result is item (5) of Theorem 4.4.

Lemma 4.11 *Let f be a homeomorphism on the unit circle S^1, and $\mu \in \mathcal{M}(S^1)$ is topologically stable for f. Then every isolated periodic point of f in the support of μ is topologically hyperbolic.*

Proof. Let μ be a topologically stable measure of a circle homeomorphism f. Suppose, on the contrary, that there is an isolated periodic point $p \in supp(\mu)$ which is not topologically hyperbolic. Let k be the period of p. Since p is not topologically hyperbolic, we have that $f^{2k}(x) - x$ does not change of sign nearby p. Moreover, since p is isolated, there is a constant $\varepsilon > 0$ such that $[p - 2\varepsilon, p + 2\varepsilon]$ is a compact interval satisfying that the graph of $f^{2k}|_{[p-2\varepsilon, p+2\varepsilon]}$ intersects the diagonal $\{(x, x) : x \in [p - 2\varepsilon, p + 2\varepsilon]\}$ only at (p, p). Then such a graph is located either on the top or on the bottom of that diagonal. From this we deduce the following alternatives:

- $d(f^{2kn}(y), p) \leq \varepsilon$ for every $(n, y) \in \mathbb{N} \times [p - \varepsilon, p]$ and $d(f^{-2kn}(y), p) \leq \varepsilon$ for every $(n, y) \in \mathbb{N} \times [p, p + \varepsilon]$;
- $d(f^{-2kn}(y), p) \leq \varepsilon$ for every $(n, y) \in \mathbb{N} \times [p - \varepsilon, p]$ and $d(f^{2kn}(y), p) \leq \varepsilon$ for every $(n, y) \in \mathbb{N} \times [p, p + \varepsilon]$.

We assume the first alternative, i.e., the graph is on the top of the diagonal. For the second alternative, the proof is similar. Take $\delta > 0$ corresponding to $\varepsilon/4$ by the topological stability of μ. By pushing the graph of f^{2k} on top of the diagonal in order to make disappear the periodic point p, we obtain a homeomorphism g with $d(f, g) \leq \delta$ such that for every $x \in [p - \varepsilon, p + \varepsilon]$ there are $n_-, n_+ \in \mathbb{N}$ satisfying $g^{-2kn_-}(x) \notin [p - 2\varepsilon, p + 2\varepsilon]$ and $g^{2kn_+}(x) \notin [p - 2\varepsilon, p + 2\varepsilon]$. In other words,

$$d(g^{-2kn_-}(x), p) \geq 2\varepsilon \quad \text{and} \quad d(g^{2kn_+}(x), p) \geq 2\varepsilon. \tag{4.9}$$

Since $d(f, g) \leq \delta$, there is an upper semicontinuous compact-valued map H of S^1 with g-invariant domain satisfying $\mu(S^1 \setminus \text{Dom}(H)) = 0$, $\mu(H(x)) = 0$ for all $x \in \text{Dom}(H)$, $d(H, Id) \leq \frac{\varepsilon}{4}$ and $f \circ H = H \circ g$. Since $\mu(S^1 \setminus \text{Dom}(H)) = 0$ and

$p \in \text{supp}(\mu)$, we can choose $x \in \text{Dom}(H)$ (i.e., $H(x) \neq \emptyset$) with $d(p, x) \leq \frac{\varepsilon}{4}$. In particular, $x \in [p - \varepsilon, p + \varepsilon]$ and so we can fix n_- and n_+ as in (4.9). Since $H(x) \neq \emptyset$, we can also fix $y \in H(x)$. Since $d(H, Id) \leq \frac{\varepsilon}{4}$, we have $d(x, y) \leq \frac{\varepsilon}{4}$, and so $y \in [p - \varepsilon, p + \varepsilon]$.

If $y \in [p - \varepsilon, p]$, then $d(f^{2kn}(y), p) \leq \varepsilon$ for every $n \in \mathbb{N}$ by the first alternative. However, $f^{2kn}(y) \in f^{2kn}(H(x)) = H(g^{2kn}(x))$ thus $d(f^{2kn}(y), g^{2kn}(x)) \leq \frac{\varepsilon}{4}$ for all $n \in \mathbb{N}$. Replacing $n = n_+$, we obtain from (4.9) that

$$\begin{aligned}
\varepsilon &\geq d\big(f^{2kn_+}(y), p\big) \\
&\geq d\big(g^{2kn_+}(x), p\big) - d\big(f^{2kn_+}(y), g^{2kn_+}(x)\big) \\
&\geq 2\varepsilon - \frac{\varepsilon}{4} = \frac{5\varepsilon}{4},
\end{aligned}$$

which is a contradiction. Similarly, if $y \in [p, p + \varepsilon]$, we can get a contradiction by replacing n by $-n$ and n_+ by $-n_-$ in the argument above. Then the proof is completed.

□

Lemma 4.12 *Let f be a homeomorphism on a compact metric space X. If $\mu \in \mathcal{M}(X)$ is topologically stable for f, then* supp(μ) *has no isolated points, and every atom of μ is neither sink nor source.*

Proof. Let μ be a topologically stable measure of f. Assume, on the contrary, that there is an isolated point $x \in \text{supp}(\mu)$, i.e., there is a constant $\varepsilon > 0$ such that $B[x, \varepsilon] = \{x\}$. Since $x \in \text{supp}(\mu)$, we have $\mu(\{x\}) = \mu(B[x, \varepsilon]) > 0$, and so x is an atom. Take $\delta > 0$ corresponding to ε by the topological stability of μ. For $f = g$, by topological stability of μ, there is an upper semicontinuous compact-valued map $H : \text{Dom}(H) \to 2^X$ with g-invariant domain such that $\mu(X \setminus \text{Dom}(H)) = 0$, $\mu(H(x)) = 0$ for all $x \in \text{Dom}(H)$, and $d(H, Id) \leq \varepsilon$. Since x is an atom and $\mu(X \setminus \text{Dom}(H)) = 0$, we get $x \in \text{Dom}(H)$. Let $y \in H(X)$. Since $d(H, Id) \leq \varepsilon$, we obtain $d(y, x) \leq \varepsilon$, and so $y \in B[x, \varepsilon] = \{x\}$. Then we get $H(x) = \{x\}$, and so

$$0 < \mu(\{x\}) \leq \mu(H(x)) = 0.$$

The contradiction shows that there is no isolated point of X in $supp(\mu)$.

Next, we prove that there is no atom of μ which is either a sink or a source. Suppose not. Then there is an atom x of μ which is a sink. If x is a source, the proof is similar. Since x is a sink, there is $\varepsilon > 0$ such that $\omega(y, f)$ is the orbit of x for every $y \in B[x, \varepsilon]$. We take $\delta > 0$ corresponding to $\varepsilon > 0$ by the topological stability of μ. For $g = f$, by definition of topological stability of μ, there is an upper semicontinuous compact-valued map H with g-invariant domain satisfying $\mu(X \setminus \text{Dom}(H)) = 0$, $\mu(H(x)) = 0$ for all $x \in \text{Dom}(H)$, $d(H, Id) \leq \varepsilon$ and $f \circ H = H \circ f$. Since x is also an atom of μ and $\mu(X \setminus \text{Dom}(H)) = 0$, we obtain that $x \in \text{Dom}(H)$. Let n be the period of x and define

$$A = \bigcup_{i=0}^{n-1} H(f^i(x)).$$

Since $x \in \mathrm{Dom}(H)$, we see that A is nonempty and compact. Since $f \circ H = H \circ f$, we also obtain that A is an invariant set of f. Moreover, since $\mu(H(x)) = 0$ for all $x \in \mathrm{Dom}(H) = 0$, we get $\mu(A) = 0$.

Let $y \in H(x)$. Since $d(H, Id) \leq \varepsilon$, we get $y \in H(x) \subset B[x, \varepsilon]$. By the choice of ε, we see that $\omega(y, f)$ is the orbit of x. Moreover, since A is compact invariant set, we get $\omega(y, f) \subset A$. Hence $\mathcal{O}_f(x) \subset A$. In particular, we have $x \in A$, and so $0 < \mu(\{x\}) \leq \mu(A) = 0$ which is a contradiction. This completes the proof.

□

We now prove the main result of this section.

Proof of Theorem 4.4. (1) Let f be a topologically stable homeomorphism on the unit circle S^1. Suppose the Dirac measure δ_p is topologically stable. Then p is neither a sink nor a source by Lemma 4.12. Moreover, since f is topologically stable, the sinks and the sources of f constitute $\Omega(f)$. Then we see that $p \notin \Omega(f)$. Conversely, let $p \notin \Omega(f)$. Since f is topologically stable, the Dirac measure δ_p is topologically stable by Corollary 4.3.

(2) Suppose that Leb is topologically stable for f and $Per(f) = \emptyset$. Then we know that f is either conjugated to an irrational rotation or a Denjoy homeomorphism. If f is conjugate to an irrational rotation, then f has no topologically stable measures by Lemma 4.9, which is a contradiction. On the other hand, if f is a Denjoy homeomorphism, the Lebesgue measure Leb is supported on the gaps of f by Lemma 4.10, which is a contradiction. This completes the proof of (2).

(3) Suppose $\Omega(f) = S^1$. Then f is conjugated to circle rotations (e.g., see Sect. 11 in [23]). By Lemma 4.9, f has no topologically stable measures. This completes the proof of (3).

(4) The proof follows from Lemma 4.10 and Example 4.3.

(5) The proof follows Lemma 4.11.

(6) Suppose that f has finitely many periodic points. If f is topologically stable, by Example 4.2, we see that the Lebesgue measure Leb is topologically stable. Conversely, suppose that Leb is topologically stable for f. By Theorem 4.4(2), we see that $Per(f) \neq \emptyset$. Since $\mathrm{supp}(Leb) = S^1$, by Theorem 4.4(5), every periodic point is topologically hyperbolic. Hence f is conjugate to a Morse–Smale diffeomorphism and so it is topologically stable (e.g., see page 76 in [33]).

□

Chapter 5
Measurable Spectral Decomposition

The purpose of this chapter is to present a measurable version of Smale's spectral decomposition theorem. We first show that there is a measure expansive homeomorphism with shadowing which does not admit the spectral decomposition. In Sect. 5.1, we introduce another type of expansive measure, which is called an N-expansive measure. It is applied to characterize N-expansive homeomorphisms. In Sect. 5.2, we introduce the notion of invariant measure 1-expansiveness for homeomorphism, and prove an extension theorem for an invariant measure 1-expansive homeomorphism on its chain recurrent set. Finally, in Sect. 5.3, we introduce a measurable version of Smale's spectral decomposition, which says that any invariant measure 1-expansive homeomorphism with shadowing admits the spectral decomposition.

5.1 N-Expansive Measures

In this section, we introduce another type of expansive measure, which is called N-expansiveness for measures. It is used to characterize N-expansive homeomorphisms. Throughout this section, we let f be a homeomorphism on a compact metric space X.

Definition 5.1 A Borel measure $\mu \in \mathcal{M}(X)$ is said to be *N-expansive* for f if there is $\varepsilon > 0$ such that for every $x \in X$ there is $F \subset \Gamma_\varepsilon^f(x)$ containing x such that the cardinality of $F \leq N$, and $\mu(\Gamma_\varepsilon^f(x) \setminus F) = 0$. Here $\varepsilon > 0$ is called an N-expansive constant of μ.

Moreover, we say that f is *invariant measure N-expansive* on an invariant subset A of X if every Borel measure $\mu \in \mathcal{M}(A, f)$ is N-expansive for $f|_A$. If f is invariant measure N-expansive on X, we say that f is *invariant measure N-expansive*.

The following theorem characterizes N-expansive homeomorphisms via the notion of N-expansive measures.

K. Lee et al., *Topological Dynamics from a Measure-Theoretic Viewpoint*, Springer Asia Pacific Mathematics Series 14, https://doi.org/10.1007/978-981-95-6390-6_5

Theorem 5.1 *Let f be a homeomorphism on a compact metric space X, and let $N \in \mathbb{N}$. Then f is N-expansive if and only if every Borel measure $\mu \in \mathcal{M}(X)$ is N-expansive for f.*

Proof. Suppose that f is N-expansive with an N-expansive constant $\varepsilon > 0$. Let $\mu \in \mathcal{M}(X)$ and $x \in X$. It is easy to see that $F = \Gamma^f_\varepsilon(x)$ has no more than N points and $\mu(\Gamma^f_\varepsilon(x) \setminus F) = 0$. Then μ is N-expansive for f.

Conversely, suppose, on the contrary, that every Borel measure $\mu \in \mathcal{M}(X)$ is N-expansive for f, and f is not N-expansive. Then for each $n \in \mathbb{N}$, there are a constant $\delta_n < \frac{1}{n}$ and $x_n \in X$ such that $\Gamma^f_{\delta_n}(x_n) \setminus \{x_n\}$ contains a subset $\hat{F}_n$ with N distinct elements. We define the sequence of Borel probability measures

$$\nu_n = \frac{1}{N} \sum_{x \in \hat{F}_n} \delta_x \quad \text{and} \quad \mu = \sum_{n=1}^{\infty} 2^{-n} \nu_n,$$

where δ_x denotes the Dirac measure centered at x. Since μ is N-expansive for f, there are a constant $\delta > 0$ and a sequence of subsets $F_n \subset \Gamma^f_\delta(x_n)$ such that $x_n \in F_n$, F_n has at most N elements and $\mu(\Gamma^f_\delta(x_n)) = \mu(F_n)$ for every $n \in \mathbb{N}$.

Take $n_0 \in \mathbb{N}$ such that $\delta_{n_0} < \delta$. Then we have

$$\begin{aligned} 0 = \mu\big(\Gamma^f_\delta(x_{n_0}) \setminus F_{n_0}\big) &\geq \mu\big(\Gamma^f_{\delta_{n_0}}(x_{n_0}) \setminus F_{n_0}\big) \\ &\geq \mu\big(\hat{F}_{n_0} \setminus F_{n_0}\big) \geq 2^{-n_0} \nu_{n_0}\big(\hat{F}_{n_0} \setminus F_{n_0}\big), \end{aligned}$$

and so $\nu_{n_0}(\hat{F}_{n_0} \setminus F_{n_0}) = 0$. It follows that $\delta_x(\hat{F}_{n_0} \setminus F_{n_0}) = 0$, and so $x \notin \hat{F}_{n_0} \setminus F_{n_0}$ for every $x \in \hat{F}_{n_0}$. This implies $\hat{F}_{n_0} \subset F_{n_0}$. Since $\hat{F}_{n_0}$ has N distinct elements and F_{n_0} has at most N elements, we obtain $\hat{F}_{n_0} = F_{n_0}$. On the other hand, since $x_{n_0} \in F_{n_0}$, we get $x_{n_0} \in \hat{F}_{n_0}$ which is a contradiction. This completes the proof.

□

It is clear that if a Borel measure $\mu \in \mathcal{M}(X)$ is $(N-1)$-expansive for a homeomorphism f on X ($N \in \mathbb{N}$), then it is N-expansive for f. However the converse does not hold as we can see in the following example.

Example 5.1 For any integer $N \geq 2$ there are a homeomorphism on a compact metric space X, and a Borel measure $\mu \in \mathcal{M}(X)$, which is N-expansive but not $(N-1)$-expansive for f.

Proof. For simplicity, we consider $N = 2$. First, we construct an example as in Theorem 3.7. More precisely, let Σ_2 be the sequence space on two symbols 0 and 1 with the metric

$$d_0(x, y) = \begin{cases} \frac{1}{2^n} \text{ if } & n = \max\{k \in \mathbb{N} | \ x_i = y_i \ \text{ for all } |i| < k\}, \\ 0 & \qquad \text{if } \ x = y, \end{cases}$$

where $x = (x_i)_{i\in\mathbb{Z}}$, $y = (y_i)_{i\in\mathbb{Z}} \in \Sigma_2$. Consider the shift map $\sigma : \Sigma_2 \to \Sigma_2$ given by

$$(\sigma(x))_i = x_{i+1}, \quad \forall x = (x)_{i\in\mathbb{Z}} \in \Sigma_2.$$

Then σ is expansive and shadowing.

For each $n \in \mathbb{N}$, choose a point $p_n \in \Sigma_2$ with the period n, and let $E = \bigcup_{n\in\mathbb{N}} \mathcal{O}_\sigma(p_n)$. Take a copy F of E such that $\Sigma_2 \cap F = \emptyset$, and let $X = \Sigma_2 \cup F$. Then F can be enumerated by a bijection $b : E \to F$ which assigns an element $\sigma^k(p_n)$ of E to an element $b(\sigma^k(p_n)) := q_{nk}$ in F, where $n \in \mathbb{N}$ and $0 \le k < n$. Consider a metric D on X defined by

$$D(x, y) = \begin{cases} d_0(x, y), & \text{if } x, y \in \Sigma_2, \\ \dfrac{1}{n} + d_0\big(x, \sigma^k(p_n)\big), & \text{if } x \in \Sigma_2,\ y = q_{nk} \in F, \\ \dfrac{1}{n} + \dfrac{1}{m} + d_0\big(\sigma^k(p_n), \sigma^l(p_m)\big), & \text{if } x = q_{nk},\ y = q_{ml} \in F. \end{cases}$$

Consider a homeomorphism f of X given by

$$f(x) = \begin{cases} \sigma(x), & \text{if } x \in \Sigma_2, \\ q_{n(k+1)}, & \text{if } x = q_{nk},\ 0 \le k < n-1, \\ q_{n0}, & \text{if } x = q_{n(n-1)}. \end{cases}$$

We note that the homeomorphism f is constructed as in the proof of Theorem 3.7. It follows that f is 2-expansive but not expansive. Since f is not expansive, by Theorem 5.1, there exists a Borel measure $\mu \in \mathcal{M}(X)$ which is not expansive. Moreover, since f is 2-expansive, we see that μ is 2-expansive for f.

□

In the following proposition, we show that if a homeomorphism f is invariant measure 1-expansive, then there exists a constant $\delta > 0$ (which is independent of the choice of Borel measures) such that $\mu(\Gamma_\delta^f(x) \setminus \{x\}) = 0$ for all $x \in X$ and $\mu \in \mathcal{M}(X, f)$. Here the constant $\delta > 0$ is called an invariant measure 1-expansive constant of f.

Proposition 5.1 *Let f be a homeomorphism on a compact metric space X. If f is invariant measure 1-expansive, then there is $\delta > 0$ such that $\mu(\Gamma_\delta^f(x) \setminus \{x\}) = 0$ for all $x \in X$ and $\mu \in \mathcal{M}(X, f)$.*

Proof. Suppose not. Then for each $n \in \mathbb{N}$, there exist $x_n \in X$ and $\mu_n \in \mathcal{M}(X, f)$ such that $\mu_n(\Gamma_{1/n}^f(x_n) \setminus \{x_n\}) > 0$. Define a Borel measure μ on X by

$$\mu(A) = \sum_{n=1}^{\infty} \frac{\mu_n(A)}{2^n}$$

for all Borel subsets A of X. It is clear that μ is an invariant measure and it is 1-expansive for f. Let $\delta > 0$ be an 1-expansive constant of μ, and choose a natural number N with $1/N < \delta$. Then we have

$$\mu(\Gamma^f_{1/N}(x_N) \setminus \{x_N\}) \geq \frac{\mu_N(\Gamma^f_{1/N}(x_N) \setminus \{x_N\})}{2^N} > 0.$$

The contradiction completes the proof.

□

We know that if a homeomorphism f is expansive then it is clearly invariant measure 1-expansive. In the following example, we see that if f is 2-expansive then it need not be invariant measure 1-expansive.

Example 5.2 Let f be the homeomorphism given in Example 5.1. Since f is 2-expansive, we see that f is measure expansive. Suppose that f is invariant measure 1-expansive. Then there exists a constant $\delta > 0$ such that $\mu(\Gamma^f_\delta(x)) = \mu(\{x\})$ for all $x \in X$ and $\mu \in \mathcal{M}(X, f)$. Take a large number $n \in \mathbb{N}$ such that

$$q_{n0} \subset \Gamma^f_\delta(p_n).$$

We define a Borel measure $\mu \in \mathcal{M}(X)$ by

$$\mu(A) = \frac{1}{n}\sum_{i=0}^{n-1} \delta_{f^i(q_{n0})}(A), \ \forall A \in \beta_X,$$

where δ_z is the Dirac measure centered at $z \in X$. Since f is invariant measure 1-expansive, we have

$$\mu(\Gamma^f_\delta(p_n) \setminus \{p_n\}) > \mu(\{q_{n0}\}) > 0,$$

which is a contradiction. Hence f is not invariant measure 1-expansive.

5.2 Invariant Measures on Chain Recurrent Sets

Let f be a homeomorphism on a compact metric space X, and let Λ be an invariant subset of X. We first observe that if f is expansive, then it is expansive on Λ, i.e., there is a constant $\delta > 0$ such that if $x, y \in \Lambda$ satisfy $d(f^n(x), f^n(y)) \leq \delta$ for all $n \in \mathbb{Z}$, then $x = y$. However, the converse is not true in general.

In this section, we prove that if a homeomorphism f is invariant measure 1-expansive on its chain recurrent set $CR(f)$, then it is invariant measure 1-expansive on X. Moreover, we construct a homeomorphism f which is invariant measure 1-expansive on its nonwandering set $\Omega(f)$, but not invariant measure 1-expansive on X.

For an invariant set $\Lambda \subset X$ and $\delta > 0$, we observe that $\Gamma_\delta^{f|_\Lambda}(x) = \Gamma_\delta^f(x) \cap \Lambda$ for all $x \in X$. We denote by $\mathcal{M}(\Lambda, f)$ the collection of all Borel measures $\mu \in \mathcal{M}(\Lambda)$ that are invariant for $f|_\Lambda$.

Theorem 5.2 *Let f be a homeomorphism on a compact metric space X. Then f is invariant measure* 1*-expansive on X if and only if it is invariant measure* 1*-expansive on its chain recurrent set $CR(f)$.*

Proof. Suppose that f is invariant measure 1-expansive on $CR(f)$. Then there exists a constant $\delta > 0$ such that

$$\mu\big((\Gamma_{2\delta}^f(x) \cap CR(f)) \setminus \{x\}\big) = 0, \ \forall x \in CR(f), \ \forall \mu \in \mathcal{M}(CR(f), f).$$

We first claim that $\mu(\Gamma_{2\delta}^f(x) \setminus \{x\}) = 0$ for all $\mu \in \mathcal{M}(X, f)$ and $x \in CR(f)$. In fact, let $\mu \in \mathcal{M}(X, f)$ and $x \in CR(f)$. We define an invariant measure $\nu \in \mathcal{M}(CR(f), f)$ by $\nu(A) = \mu(A)$ for any Borel set $A \subset CR(f)$. Since f is invariant measure 1-expansive on $CR(f)$, we have

$$\begin{aligned}
&\mu(\Gamma_{2\delta}^f(x) \setminus \{x\}) \\
&= \mu((\Gamma_{2\delta}^f(x) \cap CR(f)) \setminus \{x\}) + \mu((\Gamma_{2\delta}^f(x) \cap CR(f)^c) \setminus \{x\}) \\
&= \nu((\Gamma_{2\delta}^f(x) \cap CR(f)) \setminus \{x\}) = 0.
\end{aligned}$$

This completes the proof of the claim.

We now show that f is invariant measure 1-expansive. Suppose, on the contrary, that there are an invariant measure $\mu \in \mathcal{M}(X, f)$ and $x \in X$ such that $\mu(\Gamma_\delta^f(x) \setminus \{x\}) > 0$. By the claim above, we observe that $x \notin CR(f)$. Since $\mu(CR(f)) = 1$, there is $p \in (\Gamma_\delta^f(x) \cap CR(f)) \setminus \{x\}$. Note that $\mu(\Gamma_{2\delta}^f(p) \setminus \{p\}) = 0$ since $p \in CR(f)$. Hence we have $\mu(\Gamma_\delta^f(x) \setminus \{p\}) = 0$. It implies that $\mu(\{p\}) > 0$, and so p is a periodic point of f. Let m be the period of p. For $n \in \mathbb{Z}$, we have

$$d(f^{nm+k}(x), f^k(p)) = d(f^{nm+k}(x), f^{nm+k}(p)) < \delta, \ \forall k \in \mathbb{Z}.$$

Hence we get $f^{nm}(x) \in \Gamma_\delta^f(p)$ for all $n \in \mathbb{Z}$. Let y be a limit point of $\{f^{nm}(x)\}_{n \in \mathbb{Z}}$. Note that $y \in \Gamma_\delta^f(p)$ and $y \in \omega(x, f) \cup \alpha(x, f) \subset CR(f)$. We observe that $y = p$. Indeed, we assume that $y \neq p$. We define an invariant measure $\mu_p \in \mathcal{M}(X, f)$ by

$$\mu_p(A) = \frac{1}{m} \sum_{i=0}^{m-1} \delta_{f^i(p)}(A), \ \forall A \in \beta_X,$$

where δ_z is the Dirac measure centered at $z \in X$. Since $p \in \Gamma_\delta^f(y)$, we have

$$\mu_p\big(\Gamma_\delta^f(y) \setminus \{y\}\big) \geq \mu_p(\{p\}) > 0.$$

The contradiction shows that $y = p$. It means that $f^{nm}(x)$ converges to p as $|n| \to \infty$. It is easy to see that $x \in CR(f)$, and so we get a contraction. Therefore f is invariant measure 1-expansive.

□

In the following example, we construct a homeomorphism f which is invariant measure 1-expansive on its nonwandering set $\Omega(f)$, but not invariant measure 1-expansive on X.

Example 5.3 For each $n \in \mathbb{N}$, let S_n be the circle in $\mathbb{R}^2$ centered at $(0, \frac{1}{n})$ with radius $\frac{1}{n}$, and $X = \bigcup_{n \in \mathbb{N}} S_n$. We take a homeomorphism f on X such that the origin $\mathbf{0} = (0, 0)$ is the unique fixed point of f and $\alpha(p, f) = \omega(p, f) = \{\mathbf{0}\}$ for all $p \in X$.

Since $\Omega(f) = \{\mathbf{0}\}$, it is clear that f is invariant measure 1-expansive on $\Omega(f)$. For any $\delta > 0$, there is $p_n = (0, \frac{1}{n})$ $(n \in \mathbb{N})$ such that $\mathbf{0} \in \Gamma_\delta^f(p_n)$. Denote by $\delta_\mathbf{0}$ the Dirac measure centered at $\mathbf{0}$. We observe that $\delta_\mathbf{0} \in \mathcal{M}(X, f)$ and $\delta_\mathbf{0}(\Gamma_\delta^f(p_n) \setminus \{p_n\}) > 0$. It implies that f is not invariant measure 1-expansive on X.

5.3 Measurable Spectral Decomposition

In this section we present a measurable version of spectral decomposition theorem by Smale [36] and Bowen [10]. Before that, we first give an example to show that a measure expansive homeomorphism with shadowing does not admit the spectral decomposition.

Example 5.4 Consider 2-expansive homeomorphism f on compact metric space X as in Example 5.1. Since the homeomorphism f is constructed as in the proof of Theorem 3.7, it follows that f is also shadowing. Moreover, by the construction of f, we observe that f admits infinitely many chain components. It implies that f does not admit the spectral decomposition.

In the following theorem, we show that if a homeomorphism is invariant measure 1-expansive and shadowing on its chain recurrent set $CR(f)$ then f has the spectral decomposition.

Theorem 5.3 *Let f be a homeomorphism on a compact metric space X. Suppose that f is invariant measure 1-expansive and shadowing on its chain recurrent set $CR(f)$. Then the nonwandering set $\Omega(f)$ is decomposed as:*

(1) $\Omega(f) = \bigcup_{i=1}^n \Omega_i$, where Ω_i's are disjoint, closed and invariant, and f restricted to Ω_i is topologically transitive;
(2) for each $i \in \{1, \cdots, n\}$, there is a decomposition of Ω_i into disjoint closed sets $\Omega_i = \bigcup_{j=1}^{n_i} C_{j,i}$ such that $f(C_{j,i}) = C_{j+1,i}$ for $1 \le j < n_i$, $f(C_{n_i,i}) = C_{1,i}$, and the map $f^{n_i} : C_{j,i} \to C_{j,i}$ is topologically mixing for all $1 \le j \le n_i$.

To prove the theorem, we need some lemmas.

Lemma 5.1 *Let f be a homeomorphism on a compact metric space X. If f is invariant measure* 1*-expansive on its chain recurrent $CR(f)$. Then there is $\varepsilon > 0$ such that*

$$W^s_\varepsilon(p, f) \subset W^s(p, f) \text{ and } W^u_\varepsilon(p, f) \subset W^u(p, f), \ \forall p \in Per(f).$$

Proof. Let $\varepsilon > 0$ be a constant such that $\mu(\Gamma^f_\varepsilon(x) \setminus \{x\}) = 0$ for all $x \in X$. First, we show that $W^s_\varepsilon(p, f) \subset W^s(p, f)$ for any $p \in Per(f)$. Suppose, on the contrary, that there are $p \in Per(f)$ and $y \in W^s_\varepsilon(p, f) \setminus W^s(p, f)$. Then there are $r > 0$ and a sequence $\{n_k\}_{k\in\mathbb{N}}$ such that

$$r \le d(f^{n_k}(y), f^{n_k}(p)) \le \varepsilon \text{ for all } k \in \mathbb{N}.$$

By taking a subsequence, if necessary, we assume that $f^{n_k}(y)$ converges to $y_0 \in X$ and $f^{n_k}(p)$ converges to p_0 as $k \to \infty$. Since p is periodic, we assume that $f^{n_k}(p) = p_0$ for all k. We note that $p_0 \neq y_0$ since

$$d(p_0, y_0) = \lim_{k\to\infty} d(f^{n_k}(p), f^{n_k}(y)) \ge r.$$

Since $y \in W^s_\varepsilon(p, f)$, we have

$$d(f^i(p_0), f^i(y_0)) = \lim_{k\to\infty} d(f^i(f^{n_k}(p)), f^i(f^{n_k}(y))) \le \varepsilon, \ \forall i \in \mathbb{Z}.$$

Then $p_0 \in \Gamma^f_\varepsilon(y_0)$. Let $\mu \in \mathcal{M}(X, f)$ be an invariant Borel measure supported on the periodic orbit of p_0. Then we have

$$\mu(\Gamma^f_\varepsilon(y_0) \setminus \{y_0\}) \ge \mu(\{p_0\}) > 0.$$

The contradiction shows that $W^s_\varepsilon(p, f) \subset W^s(p, f)$ for all $p \in Per(f)$.

Similarly, we can show that $W^u_\varepsilon(p, f) \subset W^u(p, f)$, which completes the proof. □

Lemma 5.2 *Let f be a homeomorphism on a compact metric space X. Suppose that f is invariant measure* 1*-expansive and shadowing on its chain recurrent set $CR(f)$. Then we have $\Omega(f) = CR(f) = \overline{Per(f)}$.*

Proof. It suffices to prove $CR(f) \subset \overline{Per(f)}$. Let $x \in CR(f)$. To prove $x \in \overline{Per(f)}$ it is enough to show that for every $0 < \varepsilon < \frac{e}{2}$ there is $y \in Per(f)$ such that $d(y, x) < \varepsilon$, where e is an invariant measure 1-expansive constant of f.

Let $\varepsilon < e/2$, and take $\delta > 0$ such that every δ-pseudo orbit of f in $CR(f)$ is ε-shadowed by a point in X. Since $x \in CR(f)$, there exists a δ-chain $\{x_i\}_{i=0}^m$ of f from x to itself. We extend it to a δ-pseudo orbit $\xi = \{x_i\}_{i\in\mathbb{Z}}$ of f by letting $x_i = x_{k(m+1)+i}$ for all $0 \le i \le m$ and $k \in \mathbb{Z}$. Since f is shadowing on $CR(f)$, there is $y \in CR(f)$ such that $d(f^n(y), x_n) \le \varepsilon$ for all $n \in \mathbb{Z}$. Note that y can be taken in $CR(f)$. Then, for each $k \in \mathbb{N}$, we have

$$d(f^i(y), f^i(f^{k(m+1)}(y))) \leq d(f^i(y), x_i) + d(x_{k(m+1)+i}, f^i(f^{k(m+1)}(y)))$$
$$\leq \varepsilon + \varepsilon \leq e, \ \forall i \in \mathbb{Z}.$$

Hence we get $\{f^{k(m+1)}(y)\}_{k\in\mathbb{N}} \subset \Gamma_e^f(y) \cap CR(f)$. Then the closure of the orbit of y is a subset of $\bigcup_{i=0}^m \Gamma_e^f(f^i(y))$. Let μ be an invariant probability measure on the closure of the orbit of y. Then we get $\mu(\bigcup_{i=0}^m \Gamma_e^f(f^i(y))) = 1$. Then there is $1 \leq i \leq m$ such that $\mu(\Gamma_e^f(f^i(y))) > 0$. Since f is invariant measure 1-expansive, we have

$$\mu(\Gamma_e^f(f^i(y))) = \mu(\{f^i(y)\}) > 0.$$

Then $y \in Per(f)$ with $d(y, x) < \varepsilon$. This completes the proof. □

Proof of Theorem 5.3. Suppose that f is invariant measure 1-expansive and shadowing on its chain recurrent set $CR(f)$. By Theorem 5.2, it follows that f is invariant measure 1-expansive on X. By Lemma 5.2, we have

$$\Omega(f) = CR(f) = \bigcup_{\lambda\in\Lambda} \Omega_\lambda,$$

where Ω_λ's are chain components of f. By Lemma 5.1, let $\varepsilon_0 > 0$ be such that

$$W^s_{\varepsilon_0}(p) \subset W^s(p, f) \quad \text{and} \quad W^u_{\varepsilon_0}(p, f) \subset W^u(p, f) \quad \text{for all} \quad p \in Per(f).$$

By the shadowing of f on $CR(f)$, take $\delta > 0$ corresponding to ε_0. For fixed $\lambda \in \Lambda$, we will prove that Ω_λ is open in $\Omega(f)$. Let

$$U = \{y \in \Omega(f) \mid d(y, \Omega_\lambda) < \delta/2\}$$

be a neighborhood of Ω_λ in $\Omega(f)$. By Lemma 5.2, we have $\Omega(f) = \overline{Per(f)}$, and so $U \cap Per(f) \neq \emptyset$. Let $p \in U \cap Per(f)$, and $y \in \Omega_\lambda$ such that $d(y, p) < \delta/2$. We claim that $p \in \Omega_\lambda$. For any $0 < \tau < \delta/2$, let $p_\tau \in Per(f)$ be such that $d(p_\tau, y) < \tau$. It is clear that $p_\tau \sim^\tau y$.

On the other hand, let $\xi_1 = \{z_i\}_{i\in\mathbb{Z}}$ be defined by $z_i = f^i(p)$ if $i < 0$ and $z_i = f^i(p_\tau)$ if $i \geq 0$. Since

$$d(f(f^{-1}(p)), p_\tau) \leq d(p, y) + d(y, p_\tau) < \delta/2 + \tau < \delta,$$

we can check that ξ_1 is a δ-pseudo orbit of f. By the shadowing of f, there is $z \in X$ such that

$$d(f^n(z), z_n) < \varepsilon_0 \text{ for all } n \in \mathbb{Z}.$$

Then $z \in W^u_{\varepsilon_0}(p, f) \cap W^s_{\varepsilon_0}(p_\tau, f) \subset W^u(p, f) \cap W^s(p_\tau, f)$. Let $k < 0 < l$ be such that $d(f(p), f^k(z)) < \tau$ and $d(f^l(z), p_\tau) < \tau$. Then the sequence $\{x_{k-1} =$

$p, x_i = f^i(z) : k \le i \le l-1\}$ is a τ-chain of f from p to p_τ. Similarly, we can construct a τ-chain from p_τ to p, and so $p \sim^\tau p_\tau \sim^\tau y$. Since τ is arbitrary, we have $p \sim y$, which implies that $p \in \Omega_\lambda$. Consequently, we derive

$$\Omega_\lambda \supset \overline{U \cap Per(f)} \supset U \cap \overline{Per(f)} = U.$$

Then Ω_λ is open in $\Omega(f)$. Since $\Omega(f)$ is compact, $\Omega(f)$ can be expressed by a finite union of disjoint closed and invariant sets, say $\Omega(f) = \Omega_1 \cup \cdots \cup \Omega_n$.

Next, we will prove that $f|_{\Omega_i}$ is topologically transitive. Let U and V be two open sets in Ω_i. There is $0 < r < \delta/2$ such that

$$B(x, r) \cap \Omega_i \subset U \quad \text{and} \quad B(y, r) \cap \Omega_i \subset V$$

for some $x \in U$ and $y \in V$. By the shadowing of f on $CR(f)$, take $\delta_r > 0$ corresponding to r. Since $x, y \in \Omega_i$, there is a δ_r-chain $\{x_i\}_{i=0}^n$ of f from x to y. Since f is shadowing on $CR(f)$, there is $z \in CR(f)$ such that

$$d(z, x) < r < \delta/2 \quad \text{and} \quad d(f^n(z), y) < r.$$

Then we have $z \in \Omega_i$ and

$$f^n(z) \in f^n(B(x, r)) \cap B(y, r) \cap \Omega_i \subset f^n(U) \cap V \neq \emptyset.$$

Therefore $f|_{\Omega_i}$ is topologically transitive, and so it completes the proof of (a).

We now prove part (b). Fix $\Omega = \Omega_i$ for some $1 \le i \le n$. For each periodic point $p \in \Omega$, denote C_p by the closure of $W^s(p, f) \cap \Omega$ in Ω. By the shadowing of f on $CR(f)$, take $0 < \delta_1$ corresponding to $\varepsilon_1 = \min\{\varepsilon_0/2, \delta/2\}$.

Claim 1. C_p is open in Ω for any periodic point p in Ω.

Let m be the period of p, and let $V = \{y \in \Omega \mid d(y, C_p) < \delta_1/4\}$ be a neighborhood of C_p in Ω. Since $\Omega \subset \overline{Per(f)}$, we can take $q \in V \cap \Omega \cap Per(f)$ with period n. Let $x_0 \in W^s(p, f) \cap \Omega$ be such that $d(x_0, q) < \delta_1/2$. We define a δ_1-pseudo orbit by

$$y_i = \begin{cases} f^i(q) & \text{if } i < 0 \\ f^i(x_0) & \text{if } i \ge 0. \end{cases}$$

Since f is shadowing on $CR(f)$, there is $y \in \Omega(f)$ with $d(f^i(y), y_i) < \varepsilon_1$ for all $i \in \mathbb{Z}$. Then we derive the following:

- For $i = 0$, we have $d(y, x_0) < \delta/2$, and so $y \in U = \Omega$.
- For $i < 0$, we have $d(f^i(y), f^i(q)) < \varepsilon_0/2$, and so $y \in W^u_{\varepsilon_0}(q, f) \subset W^u(q, f)$.
- For $i \ge 0$, we have $d(f^i(y), f^i(x_0)) < \varepsilon_0/2$, and so $y \in W^s_{\varepsilon_0/2}(x_0)$.

Since $x_0 \in W^s(p, f)$, there is $N > 0$ such that if $i \ge N$, then $d(f^i(x_0), f^i(p)) \le \varepsilon_0/2$. Hence, for any $i \in \mathbb{N}$ with $im > N$ and any $j \ge 0$, we deduce that

$$\begin{aligned} d\big(f^j(f^{mi}(y)), f^j(p)\big) &= d\big(f^{j+mi}(y), f^{j+mi}(p)\big) \\ &\le d\big(f^{j+mi}(y), f^{j+mi}(x_0)\big) + d\big(f^{j+mi}(x_0), f^{j+mi}(p)\big) \\ &\le \varepsilon_0. \end{aligned}$$

Hence

$$f^{mi}(y) \in W^s_{\varepsilon_0}(p, f) \subset W^s(p, f), \qquad \text{for all } i > \tfrac{N}{m}.$$

For $k \in \mathbb{N}$, since $f^{-mnk}(y) \in W^s(p, f)$ and $y \in \Omega$, we have

$$f^{-mnk}(y) \in \Omega \cap W^s(p, f) \subset C_p.$$

Since $y \in W^u(q, f)$, we get

$$d(f^{-mnk}(y), q) = d(f^{-mnk}(y), f^{-mnk}(q)) \to 0 \text{ as } k \to \infty.$$

Consequently, we have $q \in C_p$, and so $V \subset C_p$. This means that C_p is open in Ω.

Claim 2. If q is a periodic point with $q \in C_p$, then $C_p = C_q$.

We let m and n be the period of p and q, respectively. First, we prove that $C_q \subset C_p$. Since $q \in C_p$, there is a sequence $\{y_n\}_{n\in\mathbb{Z}} \subset W^s(p, f) \cap \Omega$ which converges to q as n tends to infinity. Fix $x_0 \in W^s(q, f) \cap \Omega$. For each $i \in \mathbb{N}$, let $\varepsilon_i = \dfrac{\varepsilon_0}{i+1}$ and take $\delta_i > 0$ corresponding to ε_i by the shadowing of f on $CR(f)$. Let $N_i \in \mathbb{N}$ be such that

$$d(y_{N_i}, q) < \delta_i/2 \qquad \text{and} \qquad d(f^{mnN_i}(x_0), q) < \delta_i/2.$$

We define a δ_i-pseudo orbit by

$$z_j = \begin{cases} f^j(x_0), & \text{if } j \le nmN_i - 1, \\ f^{j-nmN_i}(y_{N_i}), & \text{if } j \ge nmN_i. \end{cases}$$

By the shadowing of f on $CR(f)$, there is $x_i \in \Omega(f)$ such that $d(x_i, x_0) < \varepsilon_i$. Thus x_i converges to x_0 as i tends to infinity.

On the other hand, we have $x_i \in W^s(p, f)$ for all $i \in \mathbb{N}$. Indeed, for fixed $i \in \mathbb{N}$, we choose $M \in \mathbb{N}$ such that $f^M(p) = p$ and if $j \ge M$ then $d(f^j(y_{N_i}), f^j(p)) < \varepsilon_0/2$. Hence if $j \ge M$, we have

$$\begin{aligned} d\left(f^j(f^{nmN_i}(x_i)), f^j(p)\right) &\le d\left(f^{j+nmN_i}(x_i), f^j(y_{N_i})\right) + d\left(f^j(y_{N_i}), f^j(p)\right) \\ &\le \varepsilon_i + \varepsilon_0/2 \le \varepsilon_0. \end{aligned}$$

Then we obtain

$$f^{M+nmN_i}(x_i) \in W^s_{\varepsilon_0}(f^M(p)) = W^s_{\varepsilon_0}(p) \subset W^s(p, f).$$

Thus $x_i \in W^s(p, f) \cap \Omega$, and so $x_0 \in C_p$.

Next, we prove that $C_p \subset C_q$. For this, it is sufficient to show $p \in C_q$. Suppose, on the contrary, that $p \notin C_q$, and let $d = \inf\{d(q, y) | y \in C_p \setminus C_q\} > 0$. Since $q \in C_p$, there is $x \in W^s(p, f) \cap \Omega$ with $d(x, q) < d/2$. Then $x \in C_q$. Otherwise, $x \in C_p \setminus C_q$. This is a contradiction to the definition of d. Then $f^{nmk}(x) \in C_q$ for all $k \in \mathbb{N}$. However, we have

$$d(f^{nmk}(x), p) = d(f^{nmk}(x), f^{nmk}(p)) \to 0 \text{ as } k \to \infty.$$

Since C_q is closed, we get $p \in C_q$. This is a contradiction to the assumption.

Claim 3. There is $a \in \mathbb{N}$ such that $\Omega = \bigcup_{i=0}^{a-1} C_{f^i(p)}$ and $f^a|_{C_p}$ is topologically mixing.

First, we show that there is $a \in \mathbb{N}$ such that $\Omega = \bigcup_{i=0}^{a-1} C_{f^i(p)}$. Let m be the period of p and a be the smallest positive integer such that $C_{f^a(f^i(p))} = C_{f^i(p)}$ for all $i \in \mathbb{N}$. It is clear that $\bigcup_{i=0}^{a-1} C_{f^i(p)} \subset \Omega$. Conversely, assume, on the contrary, that $\Omega \setminus \bigcup_{i=0}^{a-1} C_{f^i(p)}$ is nonempty, and so it is invariant. For two nonempty open sets C_p and $\Omega \setminus \bigcup_{i=0}^{a-1} C_{f^i(p)}$ in $\Omega(f)$, there is $l \in \mathbb{Z}$ such that

$$f^l(C_p) \cap \left(\Omega \setminus \bigcup_{i=0}^{a-1} C_{f^i(p)}\right) \neq \emptyset.$$

It contradicts the fact that $\Omega \setminus \bigcup_{i=0}^{a-1} C_{f^i(p)}$ is invariant, and then $\Omega = \bigcup_{i=0}^{a-1} C_{f^i(p)}$.

Next, we show that $f^a|_{C_p}$ is topologically mixing. Let U and V be any two nonempty open sets in C_p. Let q be a periodic point in V with the period n, and let $\eta > 0$ be such that $B(q, \eta) \cap C_p \subset V$. For $0 \leq j < n$, since U is open in $C_p = C_{f^{aj}(p)} = C_{f^{aj}(q)}$, there is $w_j \in U \cap W^s(f^{aj}(q), f)$. Then there is $M_j \in \mathbb{N}$ such that

$$d(f^i(w_j), f^i(f^{aj}(q))) < \eta \quad \text{for} \quad i \geq M_j.$$

Thus for $i > 0$ with $ani - aj \geq M_j$, we have

$$d(f^{ani-aj}(w_j), q) = d(f^{ani-aj}(w_j), f^{ani-aj}(f^{aj}(q))) < \eta.$$

Let $M \in \mathbb{N}$ be such that

$$M \geq \frac{1}{a} \max \left\{ M_j | 0 \leq j < n \right\}.$$

For $k \geq M$, we have $ak = a(ni - j) \geq M_j$ for some $i, j \in \mathbb{N}$ and $0 \leq j < n$. Then

$$d(f^{ak}(w_j), q) = d(f^{ani-aj}(w_j), q) < \eta.$$

Therefore, $f^{ka}(U) \cap V \neq \emptyset$ for all $k \geq M$, and so $f^a|_{C_p}$ is topologically mixing. □

[illegible]

Next, we prove that [illegible]

Part II
Continuous Dynamical Systems

Chapter 6
Topological Dynamics for Flows

In this chapter, we introduce the basic knowledge in continuous dynamical system (or flows) for the development of Chaps. 7–9. More precisely, we will prove two important results: stability theorem and spectral decomposition theorem for flows.

6.1 Basic Concepts

Definition 6.1 Let X be a compact metric space. A *flow* (or *continuous dynamical system*) on X is a continuous map $\phi : X \times \mathbb{R} \to X$ such that

(1) $\phi(x, 0) = x$ for $x \in X$;
(2) $\phi(\phi(x, s), t) = \phi(x, s + t)$ for $s, t \in \mathbb{R}$.

For any $x \in X$, the set $\{\phi(x, t) : t \in \mathbb{R}\}$ is called the *orbit* of ϕ through x and denoted by $\phi_{\mathbb{R}}(x)$. For any $A \subset X$ and $I \subset \mathbb{R}$, we define

$$\phi_I(A) = \{\phi_t(x) : x \in A \text{ and } t \in I\},$$

where $\phi_t(x) = \phi(x, t)$. We also write $\phi_I(x)$ instead of $\phi_I(\{x\})$ for $x \in X$.

Definition 6.2 Let ϕ be a flow on a compact metric space X.

(1) A point $x \in X$ is said to be *fixed* for ϕ if $\phi_t(x) = x$ for all $t \in \mathbb{R}$. We denote by $Fix(\phi)$ the collection of all fixed points of ϕ.
(2) A point $x \in X$ is said to be *periodic* for ϕ if $\phi_t(x) = x$ for some $t > 0$. The minimal positive number t satisfies the equality is called the *period* of x. We denote by $Per(\phi)$ the collection of all periodic points of ϕ. The orbit of $x \in Per(\phi)$ is called a *periodic orbit* of ϕ.

A subset $A \subset X$ is said to be *invariant* for ϕ (or ϕ-invariant) if $\phi_t(A) = A$ for all $t \in \mathbb{R}$. Note that any orbit of ϕ is invariant for ϕ. Moreover, A is invariant if and only

K. Lee et al., *Topological Dynamics from a Measure-Theoretic Viewpoint*, Springer Asia Pacific Mathematics Series 14, https://doi.org/10.1007/978-981-95-6390-6_6

if it is a union of orbits. For an invariant subset A of ϕ, we denote by $\phi|_A$ the flow on A given by $\phi|_A(x,t) = \phi(x,t)$ for all $x \in X$ and $t \in \mathbb{R}$.

Definition 6.3 A point $y \in X$ is said to be an *ω-limit point* (resp. *α-limit point*) of x if there exists a sequence $t_i \to +\infty$ (resp. $t_i \to -\infty$) such that $\phi_{t_i}(x) \to y$ as $i \to \infty$. Denote by $\omega(x,\phi)$ (resp. $\alpha(x,\phi)$) the set of all ω-limit points (resp. α-limit points) of x.

Theorem 6.1 *Let ϕ be a flow on a compact metric space X. For any $x \in X$, the sets $\omega(x,\phi)$ and $\alpha(x,\phi)$ are nonempty, compact and invariant.*

Definition 6.4 A point $x \in X$ is said to be *nonwandering* for ϕ if for any neighborhood U of x, there is $t \geq 1$ such that $\phi_t(U) \cap U \neq \emptyset$. The collection of all nonwandering points of ϕ is called the *nonwandering set* of ϕ, denoted by $\Omega(\phi)$.

Definition 6.5 Let ϕ be a flow on a compact metric space X.

(1) For any $\delta > 0$ and $T > 0$, a sequence $\xi = \{(x_i, t_i) : x_i \in X,\ 0 < T \leq t_i,\ a \leq i \leq b\}$ $(-\infty \leq a < b \leq \infty)$ is called a *(δ, T)-pseudo orbit* (or *(δ, T)-chain*) of ϕ if $d(\phi_{t_i}(x_i), x_{i+1}) < \delta$ for all $a \leq i \leq b$. We say that a finite sequence $\{(x_i, t_i) : 0 \leq i \leq n\}$ is an $(\delta, 1)$-chain of ϕ from x to y if it is a $(\delta, 1)$-chain satisfying $x_0 = x$ and $x_n = y$.
(2) A point $x \in X$ is said to be *chain recurrent* for ϕ if for any $\varepsilon > 0$ there exist $(\varepsilon, 1)$-chain $\{(x_i, t_i)\}_{i=0}^n$ of ϕ from x to itself. The collection of all chain recurrent points of ϕ is called the *chain recurrent set* of ϕ, denoted by $CR(\phi)$.

Remark 6.1 Let ϕ be a flow on a compact metric space X. We observe that the chain recurrent set of the restriction ϕ on its chain recurrent set $CR(\phi)$ coincides with $CR(\phi)$. More precisely, a point $x \in CR(\phi)$ if and only if for any $\varepsilon > 0$ there exists an $(\varepsilon, 1)$-chain $\{(x_i, t_i)\}_{i=0}^n \subset CR(\phi)$ from x to itself.

Theorem 6.2 *Let ϕ be a flow on a compact metric space X. Then:*

(1) $\overline{Per(\phi)} \subset \Omega(\phi) \subset CR(\phi)$;
(2) $\Omega(\phi)$ and $CR(\phi)$ are compact and invariant, but $Per(\phi)$ is not closed in general.

Denote by $\mathcal{C}$ the set of continuous maps $c : \mathbb{R} \to \mathbb{R}$ with $c(0) = 0$. We recall the classical definition of expansive flow [11].

Definition 6.6 We say that a flow ϕ on a compact metric space X is *expansive* on an invariant subset $A \subset X$ if for every $\varepsilon > 0$ there is $\delta > 0$ such that if $x, y \in A$ satisfy $d(\phi_t(x), \phi_{c(t)}(y)) \leq \delta$ for all $t \in \mathbb{R}$ and some $c \in \mathcal{C}$, then $y \in \phi_{[-\varepsilon,\varepsilon]}(x)$. The constant δ is called an *expansive constant* of ϕ with respect to ε. A flow ϕ is said to be expansive if ϕ is expansive on X.

For any $x \in X$ and a constant $\delta > 0$, we set

$$\Gamma_\delta^\phi(x) = \{y \in X | d(\phi_t(x), \phi_{c(t)}(y)) \leq \delta \text{ for some } c \in \mathcal{C} \text{ and for all } t \in \mathbb{R}\},$$

which is called the *dynamic* δ-ball of ϕ centered at x. Then we can reformulate the concept of expansiveness for flows as follows. A flow ϕ is expansive if and only if for any $\varepsilon > 0$ there exists $\delta > 0$ such that $\Gamma_\delta^\phi(x) \subset \phi_{[-\varepsilon,\varepsilon]}(x)$ for all $x \in X$.

Expansiveness has many interesting consequences about the dynamics of the flow. One of those is as follows.

Lemma 6.1 *The set of fixed points of an expansive flow ϕ on a compact metric space X is finite. Moreover, every fixed point of ϕ is an isolated point of X.*

Proof. Let $\delta > 0$ be an expansive constant of ϕ with respect to $\varepsilon = 1$, and p be a fixed point of ϕ. For $x \in B(p, \delta)$, we see that

$$d(\phi_t(p), \phi_{c(t)}(x)) = d(p, x) < \delta, \ \forall t \in \mathbb{R},$$

where $c \in \mathcal{C}$ given by $c(t) = 0$ for all $t \in \mathbb{R}$. Since ϕ is expansive, we get $x \in \phi_{[-1,1]}(p) = \{p\}$, and so $x = p$. This implies that $B(p, \delta) = \{p\}$. Hence any fixed point of ϕ is isolated point of X. Moreover, since X is compact, the set of fixed points is finite. This completes the proof. □

To give examples of expansive flows, we first recall the concept of suspension as follows. For a homeomorphism f of X and a continuous map $\tau : X \to (0, \infty)$, we let

$$X^{\tau,f} = \bigcup_{0 \leq t \leq \tau(x)} (x, t)/(x, \tau(x)) \sim (f(x), 0).$$

Then the *suspension flow* $\phi^{\tau,f}$ of f under τ is a continuous map $\phi : X^{\tau,f} \times \mathbb{R} \to X^{\tau,f}$ given by

$$\phi^{\tau,f}((x, t), s) = (x, t + s), \ \ 0 \leq t + s \leq \tau(x)$$

for $(x, t) \in X^{\tau,f}$ and $s \in \mathbb{R}$. Here the map τ is called a *height function* of the suspension flow ϕ. We define $X^f = X^{\tau,f}$ and $\phi^f = \phi^{\tau,f}$ for the height function $\tau = 1$. For any $0 \leq s, t \leq 1$ and $x, y \in X$, let

$$l((x, t), (y, t)) = (1 - t)d(x, y) + td(f(x), f(y)) \text{ and } l((x, t), (x, s)) = |t - s|.$$

For any two points $\overline{x}$, $\overline{y}$ in X^f, take a finite chain $\xi = \{\overline{x} = \overline{w}_0, \overline{w}_1, \dots, \overline{w}_{n+1} = \overline{y}\}$ from $\overline{x}$ to $\overline{y}$ in X^f such that for each i, either $\overline{w}_i$ and $\overline{w}_{i+1}$ belong to $X \times \{t\}$ for some $t \in \mathbb{R}$, or $\overline{w}_i$ and $\overline{w}_{i+1}$ are on same orbit of ϕ. Define the length of the chain ξ by $l(\xi) = \sum_{i=0}^n l(\overline{w}_i, \overline{w}_{i+1})$. Then we consider a metric d^f on X^f given by

$$d^f(\overline{x}, \overline{y}) \ = \ \text{the infimum of the lengths of all chains from } \overline{x} \text{ to } \overline{y}.$$

Theorem 6.3 *Let f be a homeomorphism on a compact metric space X. Then the suspension flow ϕ^f of f is expansive if and only if f is expansive.*

To prove Theorem 6.3, we need some notations and lemmas. Let (X, d) be a compact metric space. We define a distance d' on X with respect to d by

$$d'(x, y) = \min\{d(x, y), d(f(x), f(y))\}, \ \forall x, y \in X.$$

We observe that d' is not a metric in general. However, it has a useful property: a homeomorphism f is expansive if and only if there is $\delta > 0$ such that $\hat{\Gamma}^f_\delta(x) = \{x\}$ for all $x \in X$, where

$$\hat{\Gamma}^f_\delta(x) = \{y \in X : d'(f^n(x), f^n(y)) \le \delta, \ \forall n \in \mathbb{Z}\}. \tag{6.1}$$

Moreover, for $(x, t), (y, s) \in X^f$, we have

$$d'(x, y) \le d^f((x, t), (y, s)).$$

Lemma 6.2 *Let f be a homeomorphism on a compact metric space X. Then*

$$\Gamma^{\phi^f}_\delta(y) \subset \hat{\Gamma}^f_\delta(x) \times [0, 1], \quad \forall y = (x, t) \in X^f \text{ and } 0 < \delta < \frac{1}{4}.$$

Proof. Let $0 < \delta < \frac{1}{4}$, $y = (x, t)$ and $y_0 = (x_0, t_0)$ in X^f be such that $y_0 \in \Gamma^{\phi^f}_\delta(y)$. Then there is $c \in \mathcal{C}$ such that $d^f(\phi^f_t(y), \phi^f_{c(t)}(y_0)) \le \delta$ for every $t \in \mathbb{R}$. We divide the proof that $y_0 \in \hat{\Gamma}^f_\delta(x) \times [0, 1]$ in the following two cases.
Case 1: $t = \frac{1}{2}$. Since $\phi^f_1(y) = (x, \frac{3}{2})$, $\phi^f_{c(1)}(y_0) = (x_0, t_0 + c(1))$ and $d^f(\phi^f_1(y), \phi^f_{c(1)}(y_0)) \le \delta$, we have

$$\left|\frac{3}{2} - t_0 - c(1)\right| \le \delta < \frac{1}{4},$$

and so $0 \le t_0 + c(1) \le 1$. Hence we can write $\phi^f_{c(1)}(y_0) = (f(x_0), t_1)$ for some $0 \le t_1 \le 1$. Since $d'(f(x), f(x_0)) \le d^f(\phi^f_1(y), \phi^f(y_0)) \le \delta$, we get $d'(f(x), f(x_0)) \le \delta$. Continue this process, we obtain $d'(f^n(x), f^n(x_0)) \le \delta$ for all $n \in \mathbb{Z}$, and so $x_0 \in \hat{\Gamma}^f_\delta(x)$. Since $0 \le t_0 \le 1$ by the definition of X^f we obtain that $y_0 \in \hat{\Gamma}^f_\delta(x) \times [0, 1]$.
Case 2: $t \ne \frac{1}{2}$. There is a number r with $|r| \le \frac{1}{2}$ such that $\phi^f_r(y) = \left(x, \frac{1}{2}\right)$. Let $y' = \phi^f_r(y)$ and $y'_0 = \phi^f_{c(r)}(y_0)$. We have

$$d^f(\phi^f_t(y'), \phi_{h(t)}(y'_0)) \le \delta,$$

for all $t \in \mathbb{R}$ where $h(t) = c(t + r) - c(r)$ for all $t \in \mathbb{R}$. Moreover, since $y'_0 = (x_0, t_0 + c(r))$, we get $\left|\frac{1}{2} - t_0 - c(r)\right| \le d^f(y', y'_0) \le \delta < \frac{1}{4}$, and so $\frac{1}{4} \le t_0 + c(r) \le \frac{3}{4}$. Since $h \in \mathcal{C}$ and $y' = \left(x, \frac{1}{2}\right)$, by Case 1, we obtain that $x_0 \in \hat{\Gamma}^f_\delta(x)$, and so $y_0 \in \hat{\Gamma}^f_\delta(x) \times [0, 1]$. This completes the proof.

□

Lemma 6.3 *Let f be a homeomorphism of a compact metric space X, and ϕ^f be the suspension flow of f under the height function $\tau = 1$. Then for every $0 < \varepsilon < 1$, there is $\delta > 0$ such that if*

$$d^f(\phi_t^f(y_1), \phi_{c(t)}^f(y_2)) < \delta, \ \forall t \in \mathbb{R}$$

for some $y_1, y_2 \in X^f$ and $c \in \mathcal{C}$, then $|c(t) - t| \le \varepsilon$ for every $t \in [0, 2]$.

Proof. Suppose, on the contrary, that there is a constant $0 < \varepsilon < 1$, and for each $k \in \mathbb{N}$, there are a continuous map $c_k \in \mathcal{C}$, $y_1^k, y_2^k \in X^f$ and $t_k \in [0, 2]$ such that

$$d^f(\phi_t^f(y_1^k), \phi_{c_k(t)}^f(y_2^k)) < \frac{1}{k}, \quad \forall t \ge \mathbb{R},$$

and $|c_k(t_k) - t_k| > \varepsilon$. In particular, by letting $t = 0$, we have

$$d^f(y_1^k, y_2^k) < \frac{1}{k}, \quad \forall k \in \mathbb{N}. \tag{6.2}$$

For each $k \in \mathbb{N}$, since $c_k(0) = 0$ and $|c_k(t_k) - t_k| > \varepsilon$, there is $0 \le s_k \le t_k \le 2$ such that $|c_k(s_k) - s_k| = \varepsilon$. Without loss of generality, we may assume that $c(s_k) = s_k + \varepsilon$ for all $k \in \mathbb{N}$. Then we have

$$d^f(\phi_{s_k}^f(y_1^k), \phi_{s_k+\varepsilon}^f(y_2^k)) < \frac{1}{k}, \quad \forall k \in \mathbb{N}. \tag{6.3}$$

Since X^f is compact, we may assume that $y_1^k \to y_1$, $y_2^k \to y_2$ for some $y_1, y_2 \in X^f$, and $s_k \to s$ for some $s \in [0, 2]$. By (6.2), we first observe that $y_1 = y_2$. Moreover, by (6.3), we get

$$d^f(\phi_s^f(y_1), \phi_\varepsilon^f(\phi_s^f(y_2))) = \lim_{k\to\infty} d^f(\phi_{s_k}^f(y_1^k), \phi_{s_k+\varepsilon}^f(y_2^k)) = 0,$$

and so $\phi_s^f(y_1) = \phi_\varepsilon^f(\phi_s^f(y_2))$. It implies that ϕ has a periodic point $\phi_s^f(y_1)$ with period less than 1. The contradiction completes the proof.

□

Proof of Theorem 6.3. Let f be an expansive homeomorphism on a compact metric space (X, d). We will show that the suspension flow ϕ^f of f under the constant height function $\tau = 1$ is expansive. Define a distance $d' : X \times X \to \mathbb{R}$ by

$$d'(x_1, x_2) = \min\{d(x_1, x_2), d(f(x_1), f(x_2))\}$$

for $x_1, x_2 \in X$. Since f is expansive, there is a constant $e > 0$ such that $\hat{\Gamma}_e^f(x) = \{x\}$ for all $x \in X$, where $\hat{\Gamma}_e^f(x)$ is given in (6.1). Let $0 < \varepsilon < \frac{1}{4}$, take δ_0 corresponding to $\frac{\varepsilon}{4}$ as in Lemma 6.3, and

$$0 < \delta < \frac{1}{4}\min\{1, \varepsilon, e, \delta_0\}.$$

Now, let $y_1 = (x_1, t_1)$, $y_2 = (x_2, t_2) \in X^f$ be such that

$$d^f(\phi_t^f(y_1), \phi_{c(t)}^f(y_2)) < \delta, \quad \forall t \in \mathbb{R}, \tag{6.4}$$

for some $c \in \mathcal{C}$. We claim that $y_2 \in \phi_{[-\varepsilon,\varepsilon]}^f(y_1)$.

Since $y_2 \in \Gamma_\delta^{\phi^f}(y_1)$ and $0 < \delta < \frac{1}{4}$, by Lemma 6.2, we have $y_2 \in \hat{\Gamma}_\delta^f(x_1) \times [0, 1]$. Moreover, by the choice of $0 < \delta < e$, we get $\hat{\Gamma}_\delta^f(x_1) = \{x_1\}$, and so $y_2 \in \{x_1\} \times [0, 1]$. It implies that $x_2 = x_1$. We consider the following two cases.

Case 1: $t_1 = \frac{1}{2}$. Since $x_1 = x_2$, we have

$$d^f(y_1, y_2) = |t_2 - \frac{1}{2}|.$$

On the other hand, by letting $t = 0$ in (6.4), we get $d^f(y_1, y_2) < \delta < \varepsilon$. Then $|t_2 - \frac{1}{2}| < \varepsilon$, and so $t_2 - \frac{1}{2} \in [-\varepsilon, \varepsilon]$. It implies that

$$y_2 = \phi_{t_2-\frac{1}{2}}(y_1) \in \phi_{[-\varepsilon,\varepsilon]}^f(y_1).$$

Case 2: $t_1 \neq \frac{1}{2}$. There is $r \in \mathbb{R}$ with $|r| < \frac{1}{2}$ such that $\phi_r^f(y_1) = (y_1, \frac{1}{2})$. Let $y_1' = \phi_r^f(y_1)$ and $y_2' = \phi_{c(r)}^f(y_2)$. By (6.4), we may write

$$d^f(\phi_t^f(y_1'), \phi_{h(t)}^f(y_2')) < \delta, \quad \forall t \in \mathbb{R},$$

where $h : \mathbb{R} \to \mathbb{R}$ defined by $h(t) = c(t + r) - c(r)$ for all $t \in \mathbb{R}$. Then, by Case 1, we get $y_2' \in \phi_{[-\frac{\varepsilon}{4},\frac{\varepsilon}{4}]}^f(y_1')$, and so

$$y_2 \in \phi_{[r-c(r)-\frac{\varepsilon}{4},r-c(r)+\frac{\varepsilon}{4}]}^f(y_1).$$

By the choice of δ_0, we obtain that $|c(r) - r| \le \frac{\varepsilon}{4}$, and so $y_2 \in \phi_{[-\varepsilon,\varepsilon]}^f(y_1)$. This completes the proof.

Conversely, we prove that if ϕ^f is expansive, then so is f. Let $\varepsilon > 0$ be given. By replacing ε by $\min(\varepsilon, \frac{1}{4})$ if necessary we can assume $\varepsilon < \frac{1}{2}$. Let δ be an expansive constant of ϕ^f associated to this ε where the latter uses the suspended metric d^f. Suppose that $x_1, x_2 \in X$ satisfies $d(f^n(x_1), f^n(x_2)) < \delta$ for all $n \in \mathbb{Z}$. For these x_1, x_2 it corresponds the points $y_1 = (x_1, 0)$ and $y_2 = (x_2, 0)$ in the mapping torus X^f. By the definition of the suspended metric d^f, we have

$$\begin{aligned} d^f\big(\phi_t^f(y_1), \phi_t^f(y_2)\big) \le\ & (1 - t + [t])\, d\big(f^{[t]}(x_1), f^{[t]}(x_2)\big) \\ & + (t - [t])\, d\big(f^{[t]+1}(x_1), f^{[t]+1}(x_2)\big) \end{aligned}$$

$$\leq (1 - t + [t])\delta + (t - [t])\delta$$
$$= \delta, \qquad \forall t \in \mathbb{R}.$$

Hence $y_2 \in \phi^f_{[-\varepsilon,\varepsilon]}(y_1)$. Since $\varepsilon < \frac{1}{2}$, we obtain $y_2 \in \phi^f_{[-\frac{1}{2},\frac{1}{2}]}(y_1)$ which since $y_1 = (x_1, 0)$ and $y_2 = (x_2, 0)$ can occur only if $x_1 = x_2$ by the definition of ϕ^f. Therefore, f is expansive and the proof follows.

□

In the definition of expansive flows, the expansive constant depends on the given constant ε measuring the distance between x and y along their orbits. In [35], Ruggiero dismissed this dependence and introduced another notion of expansiveness (see also Definition 3.2.11 [23]) which is a generalization of expansiveness. In this book, we use the name *geometric expansiveness*, although alternative names have been used.

Definition 6.7 We say that ϕ is *geometric expansive* on an invariant subset $A \subset X$ if there is $\delta > 0$ such that if $x, y \in A$ satisfy $d(\phi_t(x), \phi_{c(t)}(y)) \leq \delta$ for all $t \in \mathbb{R}$ and some $c \in \mathcal{C}$, then $y \in \phi_{\mathbb{R}}(x)$. Here δ is called a geometric expansive constant of ϕ. A flow ϕ is said to be geometric expansive if ϕ is geometric expansive on X.

We can reformulate the above definition by saying that a flow ϕ is geometric expansive if and only if there is $\delta > 0$ such that $\Gamma^\phi_\delta(x) \subset \phi_{\mathbb{R}}(x)$ for all $x \in X$. It is clear by definition that any expansive flow is geometric expansive. However, the converse is not true in general. To prove it we will use the following example, which was considered in [6], to study the geometric separating flows. As a result we have that the notion of geometric expansive flows is really weaker than that of expansive flows.

Example 6.1 For each $n \in \mathbb{N}$, let $A_n = \{a_{n_0}, a_{n_1}\}$ be a set of two points in $\mathbb{R}^+$ such that

$$A_n \cap A_m = \emptyset \text{ if } m \neq n, \quad \text{and} \quad A_n \to \{0\} \text{ under the Hausdorff metric.}$$

Let $X = \{\infty\} \cup (\mathbb{Z} \times \{0\}) \cup (\bigcup_{n \in \mathbb{Z}} \{-n, \dots, n\} \times A_n)$ be a subspace of the sphere $\mathbb{R}^2 \cup \{\infty\}$, and consider a homeomorphism f on X defined by

$$f(x) = \begin{cases} \infty, & \text{if } x = \infty, \\ (n+1, 0), & \text{if } x = (n, 0), \\ (i+1, a_{n_j}), & \text{if } x = (i, a_{n_j}), \\ (-n, a_{n_1}), & \text{if } x = (n, a_{n_0}), \\ (-n, a_{n_0}), & \text{if } x = (n, a_{n_1}), \end{cases}$$

where $j = 0, 1$ and $-n \leq i < n$. Then for any $\delta > 0$, we can take $n \in \mathbb{N}$ such that $1/n < \delta$ and $x_n \in \Gamma^f_\delta(y_n)$, where $x_n = (0, a_{n_0})$ and $y_n = (0, a_{n_1})$ in X. This implies that f is not expansive. Let ϕ^f be the suspension flow of f under the constant height function $\tau = 1$. By Theorem 6.3, we see that ϕ^f is not expansive. However, we can easily see that ϕ^f is geometric expansive by our construction.

We now introduce the notion of shadowing for flows. Denote by $\mathcal{H}$ the collection of all increasing homeomorphisms $h : \mathbb{R} \to \mathbb{R}$ such that $h(0) = 0$.

Definition 6.8 We say that a flow ϕ on a compact metric space X is *shadowing* on an invariant subset A of X if for every $\varepsilon > 0$ there is $\delta > 0$ such that every $(\delta, 1)$-pseudo orbit $\xi = \{(x_i, t_i)\}_{i\in\mathbb{Z}}$ of ϕ in A is ε-shadowed by a point $x \in X$, that is, there is $h \in \mathcal{H}$ such that

$$d(\phi_{h(t)}(x), \phi_{t-T_n}(x_n)) < \varepsilon, \ \forall t \in [T_n, T_{n+1}) \text{ with } n \in \mathbb{Z},$$

where

$$T_n = \begin{cases} \sum_{i=0}^{n-1} t_i & \text{if } n > 0, \\ 0 & \text{if } n = 0, \\ -(\sum_{i=n}^{-1} t_i) & \text{if } n < 0. \end{cases}$$

Here T_n is called the nth sum associated to ξ. We say that ϕ is *shadowing* if it is shadowing on X.

For given $\delta > 0$ and $0 < T_1 < T_2$, a sequence $\{(x_i, t_i)\}_{i\in\mathbb{Z}}$ is called a (δ, T_1, T_2)-pseudo orbit of ϕ if $T_1 \le t_i \le T_2$ for all $i \in \mathbb{Z}$, and

$$d(\phi_{t_i}(x_i), x_{i+1}) \le \delta, \ \forall i \in \mathbb{Z}.$$

Proposition 6.1 *Let ϕ be a flow on a compact metric space X. Then the following are equivalent:*

(1) ϕ is shadowing;
(2) for every $\varepsilon > 0$, there are $\delta > 0$ and $T > 0$ such that every (δ, T)-pseudo orbit of ϕ is ε-shadowed;
(3) for every $\varepsilon > 0$, there are $\delta > 0$ and $T > 0$ such that every $(\delta, T, 2T)$-pseudo orbit is ε-shadowed.

Komuro [26] introduced another type of shadowing for flows, which is called strong shadowing. For $\varepsilon > 0$, we denote by $\mathcal{H}(\varepsilon)$ the collection of all increasing homeomorphisms h from $\mathbb{R}$ to itself such that $h(0) = 0$ and

$$\left| \frac{h(s) - h(t)}{s - t} - 1 \right| \le \varepsilon \text{ for } t \ne s.$$

We say that a $(\delta, 1)$-pseudo orbit of ϕ can be *strongly ε-shadowed* if there are $x \in X$ and $h \in \mathcal{H}(\varepsilon)$ such that

$$d(\phi_{h(t)}(x), \phi_{t-T_i}(x_i)) \le \varepsilon, \ \ \forall t \in [T_i, T_{i+1}) \text{ for } i \in \mathbb{Z},$$

where T_i is given as before.

Definition 6.9 We say that a flow ϕ on a compact metric space X is *strongly shadowing* if for any $\varepsilon > 0$ there is $\delta > 0$ such that any $(\delta, 1)$-pseudo orbit of ϕ is strongly ε-shadowed by a point in X.

It is clear that any strongly shadowing flow is shadowing. However, the converse does not hold in general. In the following theorem, we observe that if ϕ has no fixed points, then the notion of shadowing and strongly shadowing are equivalent (for more details, see Theorem 4 in [26]).

Theorem 6.4 *Let ϕ be a flow on a compact metric space X without fixed points. Then ϕ is shadowing if and only if it is strongly shadowing.*

The following theorem shows that strong shadowing implies that the nonwandering set coincides with the chain recurrent set, and that the restriction of the flow to this set is also strongly shadowing. The proof can be found in [26].

Theorem 6.5 *Let ϕ be a strongly shadowing flow on a compact metric space X. Then $\Omega(f) = CR(\phi)$, and its restriction $\phi|_{\Omega(\phi)}$ is strongly shadowing.*

6.2 Topological Stability

For two flows ϕ, ψ on a compact metric space X, we define the C^0-distance by

$$d_{C^0}(\phi, \psi) = \sup\{d(\phi_t(x), \psi_t(x)) : x \in X \text{ and } t \in [0, 1]\}.$$

The following is the classical definition of topologically stable flows, due to Thomas [37].

Definition 6.10 We say that a flow ϕ on a compact metric space X is *topologically stable* if for every $\varepsilon > 0$ there is $\delta > 0$ such that for any flow ψ on X with $d_{C^0}(\phi, \psi) \leq \delta$, there exist a continuous map $h : X \to X$ sending orbits of ψ into orbits of ϕ, namely

$$h(\psi_{\mathbb{R}}(x)) \subset \phi_{\mathbb{R}}(h(x)), \quad \forall x \in X.$$

The following result, known as *Thomas's stability theorem* [37], is the flow analogue of Walters's stability theorem.

Theorem 6.6 *Let ϕ be a flow on a compact metric space X. If ϕ is expansive and shadowing, then it is topologically stable.*

Note that since ϕ is expansive, by Lemma 6.1, we may assume that ϕ has no fixed points. Here let us recall a property of flows without fixed points which is useful in the next chapters.

Lemma 6.4 *Let ϕ be a flow on a compact metric space X without fixed points. Then there exists $T_0 > 0$ such that for all $0 < T < T_0$, there is $\gamma > 0$ satisfying*

$$d(\phi_T(x), x) > \gamma, \ \forall x \in X.$$

Proof. If the flow ϕ has no periodic points, we choose $T_0 = 1$; otherwise, we take

$$T_0 = \{\tau > 0 | \phi_\tau(x) = x \text{ for some } x \in X\} > 0.$$

Suppose, on the contrary, that there is $0 < T < T_0$ and a sequence $\{x_n\}_{n\in\mathbb{N}}$ in X such that

$$d(\phi_T(x_n), x_n) \leq \frac{1}{n}, \quad \forall n \in \mathbb{N}.$$

Since X is compact, we may assume that $x_n \to x$ for some $x \in X$. Then we have

$$d(\phi_T(x), x) = \lim_{n\to\infty} d(\phi_T(x_n), x_n) = 0.$$

Then $\phi_T(x) = x$ with $0 < T < T_0$, which is a contradiction.

□

Proof of Theorem 6.6. Since ϕ is expansive, by Lemma 6.1, we see that $Fix(\phi)$ consists of finitely many isolated points of X. From this, we can assume that ϕ has no fixed points of X, and ϕ is strongly shadowing.

Let $T_0 > 0$ be as in Lemma 6.4. Fix $\varepsilon > 0$ and let $e > 0$ be an expansive constant of ϕ with respect to ε. For $0 < \varepsilon' < \frac{1}{8}\min\{\varepsilon, e, T_0\}$, we take $0 < \delta < \varepsilon'$ corresponding to ε' by the strong shadowing of ϕ. Let ψ be a flow with $d_{C^0}(\phi, \psi) < \delta$.

For every $y \in X$ we define A_y equal to the set of $x \in X$ such that for all $\eta, T > 0$ there is $\alpha \in \mathcal{H}(\varepsilon')$ satisfying

$$d(\phi_{\alpha(t)}(x), \psi_t(y)) \leq 2\varepsilon' + \eta, \quad \forall t \in [-T, T]. \tag{6.5}$$

At first glance if we replace $t = 0$ above one gets

$$A_y \subset B[y, 2\varepsilon'], \quad \forall y \in X. \tag{6.6}$$

Let us prove more properties for this set ($\forall y \in X$).

(i) $A_y \neq \emptyset$.

Consider the sequence $\{(\psi_i(y), t_i)\}_{i\in\mathbb{Z}}$ with $t_i = 1$ for every $i \in \mathbb{Z}$. Since $t_i \geq 1$ and

$$d(\phi_{t_i}(\psi_i(y)), \psi_{i+1}(y)) = d(\phi_1(\psi_i(y)), \psi_1(\psi_i(y))) \leq d_{C^0}(\phi, \psi) < \delta, \quad \forall i \in \mathbb{Z},$$

we get $(\psi_i(y), t_i)_{i\in\mathbb{Z}}$ is a $(\delta, 1)$-pseudo orbit. Thus, by the strong shadowing of ϕ, there are $x \in X$ and $\alpha \in \mathcal{H}(\varepsilon')$ such that

$$d(\phi_{\alpha(t)}(x), \phi_{t-T_i}(\psi_i(y))) < \varepsilon', \quad \forall t \in [T_i, T_{i+1}), \ i \in \mathbb{Z}.$$

Since $t_i = 1$ for all $i \in \mathbb{Z}$, by definition, we have $T_i = i$ for all $i \in \mathbb{Z}$, and so $T_i = [t]$, which is the largest integer less than t, for all $t \in [T_i, T_{i+1})$ and $i \in \mathbb{Z}$. We conclude that

$$d(\phi_{\alpha(t)}(x), \phi_{t-[t]}(\psi_{[t]}(y))) < \varepsilon', \quad \forall t \in \mathbb{R}.$$

Since $t - [t] \in [0, 1)$ for all $t \in \mathbb{R}$, we obtain that

$$\begin{aligned} d(\phi_{\alpha(t)}(x), \psi_t(y)) &\le d\Big(\phi_{\alpha(t)}(x), \phi_{t-[t]}(\psi_{[t]}(y))\Big) \\ &\quad + d\Big(\phi_{t-[t]}(\psi_{[t]}(y)), \psi_{t-[t]}(\psi_{[t]}(y))\Big) \\ &< \varepsilon' + d_{C^0}(\phi, \psi) \\ &< \varepsilon' + \delta < 2\varepsilon', \end{aligned}$$

which proves (6.5). Then, $x \in A_y$, and so $A_y \neq \emptyset$.

(ii) $A_y \subset \phi_{[-\varepsilon,\varepsilon]}(x)$ *for all* $x \in A_y$.

Let $x, x' \in A_y$. For $N \in \mathbb{N}$, there are sequences $\alpha_N, \alpha'_N \in \mathcal{H}(\varepsilon')$ such that for every $N \in \mathbb{N}$

$$\max\{d(\phi_{\alpha_N(t)}(x), \psi_t(y)), d(\phi_{\alpha'_N(t)}(x'), \psi_t(y))\} \le 2\varepsilon' + \frac{1}{2N}, \quad \forall t \in [-N, N].$$

Then, by the triangle inequality, we have

$$d(\phi_{\alpha_N(t)}(x), \phi_{\alpha'_N(t)}(x')) \le 4\varepsilon' + \frac{1}{N}, \quad \forall N \in \mathbb{N}, t \in [-N, N].$$

It is clear that $\{\alpha_N\}_{N\in\mathbb{N}}$ and $\{\alpha'_N\}_{N\in\mathbb{N}}$ both are pointwise bounded and equicontinuous. By the Arzela–Ascoli theorem, there are subsequences of $\{\alpha_N\}_{N\in\mathbb{N}}$ and $\{\alpha'_N\}_{N\in\mathbb{N}}$, we still denote them by $\{\alpha_N\}_{N\in\mathbb{N}}$ and $\{\alpha'_N\}_{N\in\mathbb{N}}$ respectively, such that $\{\alpha_N\}_{N\in\mathbb{N}}$ converges compactly to α, and $\{\alpha'_N\}_{N\subset\mathbb{N}}$ converges compactly to α' for some $\alpha, \alpha' \in \mathcal{H}(\varepsilon')$. For any fixed $t \in \mathbb{R}$, take $N > 0$ such that $t \in [-N, N]$. Then we have

$$d(\phi_{\alpha(t)}(x), \phi_{\alpha'(t)}(x')) = \lim_{N\to\infty} d(\phi_{\alpha_N(t)}(x), \phi_{\alpha'_N(t)}(x')) \le 4\varepsilon'.$$

Define a homeomorphism $\beta : \mathbb{R} \to \mathbb{R}$ by $\beta(t) = \alpha' \circ \alpha^{-1}(t)$ for $t \in \mathbb{R}$. Then we see that $\beta \in \mathcal{C}$ and

$$d(\phi_t(x), \phi_{\beta(t)}(x')) < 4\varepsilon' < e, \quad \forall t \in \mathbb{R}.$$

Since ϕ is expansive, we obtain that $x' \in \phi_{[-\varepsilon,\varepsilon]}(x)$. Since $x' \in A_y$ is arbitrary, $A_y \subset \phi_{[-\varepsilon,\varepsilon]}(x)$.

(iii) *If* $y_n \to y$ *as* $n \to \infty$ *and* $x_n \in A_{y_n} \to x$, *then* $x \in A_y$. *In particular* A_y *is closed.*

For every $n \in \mathbb{N}$, by definition of A_{y_n}, there is a sequence $\{\alpha_n\}_{n\in\mathbb{N}} \in \mathcal{H}(\varepsilon')$ such that

$$d(\phi_{\alpha_n(t)}(x_n), \psi_t(y_n)) \le 2\varepsilon' + \frac{1}{n}, \quad \forall t \in [-n, n],\ n \in \mathbb{N}.$$

Since $\{\alpha_n\}_{n\in\mathbb{N}}$ is pointwise bounded and equicontinuous, by the Arzela–Ascoli theorem, there is a subsequence of $\{\alpha_n\}_{n\in\mathbb{N}}$, we still denote it by $\{\alpha_n\}_{n\in\mathbb{N}}$, such that $\{\alpha_n\}_{n\in\mathbb{N}}$ converges compactly to α for some $\alpha \in \mathcal{H}(\varepsilon')$. For any fixed $t \in \mathbb{R}$, take $n \in \mathbb{N}$ such that $t \in [-n, n]$, we have

$$d(\phi_{\alpha(t)}(x), \psi_t(y)) = \lim_{n\to\infty} d(\phi_{\alpha_n(t)}(x_n), \psi_{t_n}(y_n)) \le 2\varepsilon'.$$

This implies $x \in A_y$. In particular, we see that A_y is closed by applying the above argument with $y_n = y$.

(iv) $A_{\psi_s(y)} \subset \phi_{\mathbb{R}}(x)$ *for all* $x \in A_y$ *and* $s \in \mathbb{R}$.
Let $\hat{x} \in A_{\psi_s(y)}$. Then, there is $\hat{\alpha} \in \mathcal{H}(\varepsilon')$ such that

$$d(\phi_{\hat{\alpha}(t)}(\hat{x}), \psi_{t+s}(y)) \le 2\varepsilon', \quad \forall t \in \mathbb{R}.$$

Since $x \in A_y$ there is $\alpha \in \mathcal{H}(\varepsilon')$ such that

$$d(\phi_{\alpha(t)}(x), \psi_t(y)) \le 2\varepsilon', \quad \forall t \in \mathbb{R}.$$

Replacing t by $t + s$ yields

$$d(\phi_{\alpha(t+s)}(x), \psi_{t+s}(y)) \le 2\varepsilon', \quad \forall t \in \mathbb{R}.$$

Then, by the triangle inequality, we get

$$d(\phi_{\hat{\alpha}(t)}(\hat{x}), \phi_{\alpha(t+s)}(x)) \le 4\varepsilon', \quad \forall t \in \mathbb{R}.$$

Define a homeomorphism $\beta : \mathbb{R} \to \mathbb{R}$ by $\beta(t) = \alpha(t + s) - \alpha(s)$ for $t \in \mathbb{R}$. We see that $\beta \in \mathcal{C}$ and

$$d(\phi_{\hat{\alpha}(t)}(\hat{x}), \phi_{\beta(t)}(\phi_{\alpha(s)}(x)) \le 4\varepsilon' < e, \quad \forall t \in \mathbb{R}.$$

Since ϕ is expansive, we get

$$\hat{x} \in \phi_{[\alpha(s)-\varepsilon,\alpha(s)+\varepsilon]}(x) \subset \phi_{\mathbb{R}}(x).$$

This completes the proof of (iv).

Denote by 2^X the set of nonempty compact subsets of X. We know that 2^X is a compact metric space if it endowed with the *Hausdorff distance* defined by

$$d_H(K, K') = \inf\{\rho > 0 : K \subset B(K', \rho) \text{ and } K' \subset B(K, \rho)\},$$

where

$$B(L,\rho) = \{z \in X : d(z,l) < \rho \text{ for some } l \in L\}$$

is the ρ-ball around $L \in 2^X$. By (iii) we have a map $A : X \to 2^X$ that takes $y \in X$ to A_y.

(v) *The set-valued map A is continuous with respect to the Hausdorff metric.*

Given $y \in X$ and $\eta, T > 0$ we define

$$A_{y,\eta,T} := \{x \in X : \exists \alpha \in \mathcal{H}(\varepsilon') \text{ s.t. } d(\phi_{\alpha(t)}(x), \psi_t(y)) \leq 2\varepsilon' + \eta, \ \forall t \in [-T, T]\}.$$

From the arguments above we see that $A_{y,\eta,T}$ is compact for all $\eta, T > 0$ and $y \in X$.

Moreover, for a given $y \in X$, we observe that

(a) if $0 < \eta_2 \leq \eta_1$, then $A_{y,\eta_2,T} \subset A_{y,\eta_1,T}$ for all $T > 0$;
(b) if $0 < T_1 \leq T_2$, then $A_{y,\eta,T_2} \subset A_{y,\eta,T_1}$ for all $\eta > 0$;
(c) if $\eta_i \to 0$ and $T_i \to \infty$, then

$$A_y = \bigcap_{i \in \mathbb{N}} A_{y,\eta_i,T_i}, \quad \forall y \in X.$$

We will prove the following property: for every $\lambda > 0$, there are $\eta, T > 0$ such that

$$A_{y',\eta,T} \subset B(A_{y'}, \lambda), \quad \forall y' \in X. \qquad (P)$$

In fact, let $\lambda > 0$. It follows from (c) and the compactness of A_y and $A_{y,\eta,T}$ that for every $y \in X$ there are constants $\eta_y, T_y > 0$ such that

$$A_{y,\eta_y,T_y} \subset B(A_y, \frac{\lambda}{2}). \tag{6.7}$$

Since ψ is continuous, for every $y \in X$ there is a neighborhood U_y such that

$$d(\psi_t(y), \psi_t(y')) \leq \frac{\eta_y}{2}, \quad \forall t \in [-T_y, T_y] \text{ and } \forall y' \in U_y.$$

Let $y' \in U_y$ and $x' \in A_{y',\frac{\eta_y}{2},T_y}$. By definition there is $\alpha \in \mathcal{H}(\varepsilon')$ such that

$$d(\phi_{\alpha(t)}(x'), \psi_t(y')) \leq 2\varepsilon' + \frac{\eta_y}{2}, \quad \forall t \in [-T_y, T_y].$$

Then, by the triangle inequality, we get

$$d(\phi_{\alpha(t)}(x'), \psi_t(y)) \leq 2\varepsilon' + \eta_y, \quad \forall t \in [-T_y, T_y].$$

Consequently, we obtain that

$$A_{y',\frac{\eta_y}{2},T_y} \subset A_{y,\eta_y,T_y}, \quad \forall y' \in U_y.$$

Now, since X is compact, there are $y_1, \cdots, y_k \in X$ such that $U_{y_1}, \cdots, U_{y_k}$ is an open covering of X.

We will prove that

$$T = \max\{T_{y_1}, \cdots, T_{y_k}\} \quad \text{and} \quad \eta = \frac{1}{2}\min\{\eta_{y_1}, \cdots, \eta_{y_k}\}$$

satisfies the requirements in (P).

Indeed, let $y' \in X$, and an open set U_{y_j} such that $y' \in U_{y_j}$ for some $j = 1, \cdots, k$. It follows that

$$A_{y',\eta,T} \subset A_{y',\frac{\eta_{y_j}}{2},T_{y_j}} \subset A_{y_j,\eta_{y_j},T_{y_j}} \overset{(6.7)}{\subset} B(A_{y_j}, \frac{\lambda}{2}).$$

Fix $z \in A_{y',\frac{\eta}{2},T}$. We see that $d(z, A_{y_j}) < \frac{\lambda}{2}$. On the other hand, since

$$A_{y'} \subset A_{y',\frac{\eta}{2},T} \subset B(A_{y_j}, \frac{\lambda}{2}),$$

we have $d_H(A_{y'}, A_{y_j}) < \frac{\lambda}{2}$. Consequently,

$$d(z, A_{y'}) \leq d(z, A_{y_j}) + d(A_{y_j}, A_{y'}) < \lambda$$

so we have $z \in B(A_{y'}, \lambda)$. Hence we get

$$A_{y',\eta,T} \subset B(A_{y'}, \lambda),$$

which proves (P).

Finally, let $\lambda > 0$, and $\{y_n\}_{n\in\mathbb{N}}$ be a sequence in X which converges to $y \in X$. By (iii), there is $N_1 \in \mathbb{N}$ such that

$$A_{y_n} \subset B(A_y, \lambda), \quad \forall n \geq N_1. \tag{6.8}$$

On the other hand, by (P), we can fix $\eta, T > 0$ such that

$$A_{y_n,\eta,T} \subset B(A_{y_n}, \lambda), \quad \forall n \in \mathbb{N}. \tag{6.9}$$

Next, we claim that there is $N_2 \in \mathbb{N}$ such that

$$A_y \subset A_{y_n,\eta,T}, \quad \forall n \geq N_2. \tag{6.10}$$

Take $x \in A_{y,\frac{\eta}{2},T}$. By definition there is $\alpha \in \mathcal{H}(\varepsilon')$ such that

$$d(\phi_{\alpha(t)}(x), \psi_t(y)) \leq 2\varepsilon' + \frac{\eta}{2}, \quad \forall t \in [-T, T].$$

Since $y_n \to y$ as $n \to \infty$, there is $N_2 \in \mathbb{N}$ such that for any $n \geq N_2$,

$$d(\psi_t(y), \psi_t(y_n)) \leq \frac{\eta}{2}, \quad \forall t \in [-T, T].$$

By the triangle inequality, we have

$$d(\phi_{\alpha(t)}(x), \psi_t(y_n)) \leq 2\varepsilon' + \eta, \quad \forall t \in [-T, T] \text{ and } \forall n \geq N_2,$$

and so $x \in A_{y_n,\eta,T}$. Then we get

$$A_{y,\frac{\eta}{2},T} \subset A_{y_n,\eta,T}, \quad \forall n \geq N_2.$$

Since $A_y \subset A_{y,\frac{\eta}{2},T}$, by Item (c), we obtain (6.10).

Combining (6.9) and (6.10), we see that $A_y \subset B(A_{y_n}, \lambda)$ for all $n \geq N_2$. Let $N = \max\{N_1, N_2\}$. By (6.8), we obtain that

$$D(A_y, A_{y_n}) \leq \lambda, \quad \forall n \geq N.$$

Since $\lambda > 0$ is arbitrary, we get $\lim_{n\to\infty} D(A_y, A_{y_n}) = 0$, which proves (v).

We say that $h : X \to X$ is a *selection* of A if $h(y) \in A_y$ for every $y \in X$. Selections exist by the axiom of choice. Since $2\varepsilon' < \varepsilon$, it follows from (6.6) that

$$d_{C^0}(h, id_X) \leq \varepsilon$$

for all selections h of A.

Moreover, every selection of A sends orbits of ψ into orbits of ϕ. Indeed, let h be a selection and $y \in X$. By taking $x = h(y)$ in (iv) we get

$$h(\psi_s(y)) \in A_{\psi_s(y)} \subset \phi_{\mathbb{R}}(x) = \phi_{\mathbb{R}}(h(y)) \quad (\forall s \in \mathbb{R}),$$

which proves the assertion.

Then, we are left to find a *continuous* selection. This is done as follows: Let $y \in X$. We can choose $x \in A_y$ by (i). We observe that A_y is closed by (iii), and $A_y \subset \phi_{[-\varepsilon,\varepsilon]}(x)$ by (ii). Since ϕ has no closed orbits of small period, the map

$$t \in [-\varepsilon, \varepsilon] \mapsto \phi_t(x) \in \phi_{[-\varepsilon,\varepsilon]}(x)$$

is a homeomorphism. From this we deduce that there is a unique $h_*(y) \in A_y$ such that for every $z \in A_y$, there is $w \geq 0$ such that $\phi_w(z) = h_*(y)$. This defines a selection h_* of H. By (v), we obtain that h_* is continuous. This completes the proof.

□

6.3 Spectral Decomposition

In the last section of this chapter, we present another fundamental result in topological dynamics, which is known as the spectral decomposition theorem. We first give the definitions of topological transitivity for flows.

Definition 6.11 Let ϕ be a flow on a compact metric space X. We say that ϕ is *topologically transitive* on an invariant set $A \subset X$ if for any pair of nonempty open sets U and V in A, there is an integer $T \geq 1$ such that $\phi_T(U) \cap V \neq \emptyset$.

Definition 6.12 We say that ϕ has the *spectral decomposition* if the nonwandering set $\Omega(\phi)$ of ϕ is decomposed as the finite number of basic sets $\Omega_i \subset \Omega(\phi)$, i.e.,

$$\Omega(\phi) = \bigcup_{i=1}^{m} \Omega_i,$$

where Ω_i's are disjoint, closed and invariant, and ϕ is topologically transitive on Ω_i for $1 \leq i \leq m$.

In this direction, Komuro [26] present a topological version of the spectral decomposition theorem for expansive flows with shadowing on compact metric spaces.

Theorem 6.7 *Let ϕ be a flow on a compact metric space X. Suppose that ϕ is expansive and shadowing on its chain recurrent set $CR(\phi)$. Then the nonwandering set $\Omega(\phi)$ of ϕ is decomposed as*

$$\Omega(f) = \bigcup_{i=1}^{m} \Omega_i,$$

where Ω_i's are disjoint, closed and invariant, and ϕ is topologically transitive on Ω_i for $1 \leq i \leq m$.

To prove the theorem, we need some definitions and lemmas. For any $x, y \in X$ and $\delta > 0$, we say that $x \sim^{\delta} y$ if there are a $(\delta, 1)$-chain from x to y and a $(\delta, 1)$-chain from y to x. If we define $x \sim y$ whenever $x \sim^{\delta} y$ for any $\delta > 0$, then we see that $\sim$ gives an equivalence relation on the set $CR(\phi)$. An equivalence class of $\sim$ is called a *chain component* of ϕ.

Lemma 6.5 *Let ϕ be a flow on a compact metric space X. If $d(x, y) < \delta$ for $x, y \in CR(\phi)$ and $\delta > 0$, then $x \sim^{\delta} y$.*

Proof. Let $\delta > 0$, and $x, y \in CR(\phi)$ be such that $d(x, y) < \delta$. Take $\varepsilon = \delta - d(x, y)$. Since $x \in CR(\phi)$, there is an $(\varepsilon, 1)$-chain $\{(x_0, t_0), \cdots, (x_n, t_n)\}$ from x to itself. Since

$$d(\phi_{t_{n-1}}(x_{n-1}), y) \leq d(\phi_{t_{n-1}}(x_{n-1}), x) + d(x, y) < \delta,$$

we see that $\{(x_0, t_0), \cdots, (x_{n-1}, t_{n-1}), (y, t_n)\}$ is a $(\delta, 1)$-chain from x to y. Similarly, we can construct a $(\delta, 1)$-chain from y to x. Hence we obtain $x \sim^{\delta} y$.

□

Lemma 6.6 *Let ϕ be a flow on a compact metric space X, and let $\varepsilon > 0$. Suppose that ϕ is expansive with an expansive constant $e > 0$ with respect to $\varepsilon > 0$. If, for some $x, y \in X$ and $g \in \mathcal{H}(e)$,*

$$d(\phi_t(x), \phi_{g(t)}(y)) < e, \quad \forall t > 0 \ \textit{[resp. } t < 0\textit{]},$$

then

$$d(\phi_t(x), \phi_{g(t)}\phi_{[-\varepsilon,\varepsilon]}(y)) \to 0 \quad \textit{as } t \to \infty \quad \textit{[resp. } t \to -\infty\textit{]}.$$

Proof. Suppose that

$$d(\phi_t(x), \phi_{g(t)}(y)) \le e, \qquad \text{for all } t > 0,$$

for some $x, y \in X$ and $g \in \mathcal{H}(e)$. Assume, on the contrary, that there exist a sequence $\{t_n\}_{n\in\mathbb{N}}$ in $\mathbb{R}$ with $t_n \to \infty$ as $n \to \infty$, and a constant $r > 0$ such that

$$d(\phi_{t_n}(x), \phi_{g(t_n)}\phi_{[-\varepsilon,\varepsilon]}(y)) > r, \quad \forall n \in \mathbb{N}.$$

Since X is compact, we may assume that $\phi_{t_n}(x) \to x_0$ and $\phi_{g(t_n)}(y) \to y_0$ as $n \to \infty$ for some $x_0, y_0 \in X$. Then $d(x_0, \phi_{[-\varepsilon,\varepsilon]}(y_0)) \ge r$. For each $n \in \mathbb{N}$, define a homeomorphism $h_n \in \mathcal{H}(e)$ by

$$h_n(t) = g(t_n + t) - g(t_n), \quad \forall t \in \mathbb{R}.$$

The family $\{h_n\}_{n\in\mathbb{N}}$ is pointwise bounded and equicontinuous. By the Arzela–Ascoli theorem, there exists a subsequence $\{h_{n_k}\}_{k\in\mathbb{N}}$ of $\{h_n\}_{n\in\mathbb{N}}$ converging compactly to a homeomorphism $h : \mathbb{R} \to \mathbb{R}$. Clearly, $h \in \mathcal{H}(e)$. Fix $t \in \mathbb{R}$. Choose $N > 0$ and $k_0 \in \mathbb{N}$ such that $t \in [-N, N]$ and $t + t_{n_k} > 0$ for $k \ge k_0$. Then, for any $k \ge k_0$,

$$\begin{aligned} d(\phi_t(x_0), \phi_{h(t)}(y_0)) &= \lim_{k\to\infty} d(\phi_t(\phi_{t_{n_k}}(x)), \phi_{h_{n_k}(t)}(\phi_{g(t_{n_k})}(y))) \\ &= \lim_{k\to\infty} d(\phi_{t+t_{n_k}}(x), \phi_{g(t+t_{n_k})}(y)) \le e. \end{aligned}$$

By the expansiveness of ϕ, it follows that $y_0 \in \phi_{[-\varepsilon,\varepsilon]}(x_0)$, which is a contradiction.

Similarly, we can show that if

$$d(\phi_t(x), \phi_{g(t)}(y)) < e, \quad \forall t < 0,$$

for some $x, y \in X$ and $g \in \mathcal{H}(e)$, then

$$d(\phi_t(x), \phi_{g(t)}\phi_{[-\varepsilon,\varepsilon]}(y)) \longrightarrow 0 \quad \text{as } t \to -\infty.$$

This completes the proof.

□

Lemma 6.7 *Let ϕ be a flow on a compact metric space X. Suppose that ϕ is expansive and shadowing on its chain recurrent set $CR(\phi)$. Then we have $CR(\phi) = \overline{Per(\phi)}$.*

Proof. Let $e > 0$ be an expansive constant with respect to 1. For $0 < \varepsilon < \min\{e/4, 1\}$, choose a constant $\delta > 0$ corresponding to ε by the strong shadowing of $\phi|_{CR(\phi)}$. Let $x \in CR(\phi)$, and take a $(\delta, 1)$-chain $\{(x_i, t_i)\}_{i=0}^{m}$ in $CR(\phi)$ from x to itself. We extend it to a periodic $(\delta, 1)$-pseudo orbit $\xi = \{(x_i, t_i)\}_{i\in\mathbb{Z}}$ through x by letting $x_i = x_{i+km}$ and $t_i = t_{i+km}$ for all $k \in \mathbb{Z}$ and $0 \le i \le m-1$. By the strong shadowing of ϕ on $CR(\phi)$, there exist $p \in X$ and $h \in \mathcal{H}(\varepsilon)$ such that

$$d(\phi_{h(t)}(p), \phi_{t-T_i}(x_i)) < \varepsilon, \quad \forall t \in [T_i, T_{i+1}),$$

where T_i is i^{th} sum associated to ξ. Let $T = \sum_{i=0}^{m-1} t_i$. Then, for any $k \in \mathbb{Z}$, we have

$$\begin{aligned} &d(\phi_{h(t)}(p), \phi_{h(t+kT)}(p)) \\ &\le d(\phi_{h(t)}(p), \phi_{t-T_i}(x_i)) + d(\phi_{t-T_i}(x_i), \phi_{h(t+kT)}(p)) < 2\varepsilon, \quad \forall t \in \mathbb{R}. \end{aligned} \tag{6.11}$$

Let $g : \mathbb{R} \to \mathbb{R}$ be a homeomorphism given by

$$g(t) = h(t + kT) - h(kT), \quad \forall t \in \mathbb{R}.$$

For $t \in \mathbb{R}$, set $s = h^{-1}(t)$. Then, by (6.11), we have

$$d(\phi_t(p), \phi_{g\circ h^{-1}(t)}(\phi_{h(kT)}(p)) = d(\phi_{h(s)}(p), \phi_{h(s+kT)}(p)) < 2\varepsilon.$$

Hence,

$$\phi_{h(kT)}(p) \in \Gamma^{\phi}_{2\varepsilon}(p), \quad \text{for all } k \in \mathbb{Z}.$$

Since ϕ is expansive, it follows that $\phi_{h(kT)}(p) \in \phi_{[-1,1]}(p)$ for all $k \in \mathbb{Z}$. Since $h(kT) \to \infty$ as $k \to \infty$, we deduce that p is a periodic point of ϕ. Since $d(x, p) < \varepsilon$ and ε is arbitrary, we conclude that $x \in \overline{Per(\phi)}$. This completes the proof.

□

Proof of Theorem 6.7. Suppose ϕ is expansive and shadowing on its chain recurrent set $CR(\phi)$. By Lemma 6.1, every fixed points of ϕ is isolated from $CR(\phi)$. Combining this with Theorem 6.4, we may assume that ϕ has no fixed points and the restriction $\phi|_{CR(\phi)}$ is strongly shadowing. By Lemma 6.7, it follows that $CR(\phi) = \Omega(\phi) = \overline{Per(\phi)}$. Hence, the nonwandering set $\Omega(\phi)$ can be expressed as the union of its chain components

$$\Omega(\phi) = \bigcup_{\lambda\in\Lambda} \Omega_\lambda,$$

where each Ω_λ is a chain components of ϕ. Note that every Ω_λ is invariant and closed in $\Omega(\phi)$.

We first show that each Ω_λ is open in $\Omega(\phi)$. Let $e > 0$ be an expansive constant of ϕ with respect to 1. By the strong shadowing of $\phi|_{CR(\phi)}$, choose a constant $\delta > 0$ such that every $(\delta, 1)$-pseudo orbit of ϕ in $\Omega(\phi)$ can be strongly shadowed by a point in $\Omega(\phi)$. For each $\lambda \in \Lambda$, define a neighborhood U_λ of Ω_λ in $\Omega(\phi)$ by

$$U_\lambda = \{x \in \Omega(\phi) | d(x, \Omega_\lambda) < \delta/2\}.$$

We claim that $p \in \Omega_\lambda$ for every periodic point $p \in Per(\phi) \cap U_\lambda$. Indeed, take $y \in \Omega_\lambda$ such that $d(y, p) < \delta/2$. For any $0 < \varepsilon < \delta/2$, by Lemma 6.7, there exists a periodic point p_ε such that $d(y, p_\varepsilon) < \varepsilon$. Then, by Lemma 6.5, we have $p_\varepsilon \sim^\varepsilon y$.

To show that $p_\varepsilon \sim^\varepsilon p$, take two constants $T, S > 1$ such that $\phi_T(p) = p$ and $\phi_S(p_\varepsilon) = p_\varepsilon$. Consider a sequence $\xi = \{(x_i, t_i)\}_{i\in\mathbb{Z}}$ defined by

$$(x_i, t_i) = \begin{cases} (\phi_{iT}(p),\ T), & i \geq 0, \\ (\phi_{iS}(p_\varepsilon),\ S), & i < 0. \end{cases}$$

Since

$$d(p, p_\varepsilon) \leq d(p, y) + d(y, p_\varepsilon) \leq \delta,$$

it follows that ξ is a $(\delta, 1)$-pseudo orbit contained in $\Omega(\phi)$. By the strong shadowing of $\phi|_{CR(\phi)}$, there exist $x \in X$ and $g \in \mathcal{H}(e)$ such that

$$d(\phi_{g(t)}(x), \phi_{t-T_i}(x_i)) < e, \text{ for all } T_i \leq t < T_{i+1} \text{ with } i \in \mathbb{Z},$$

where T_i is the ith sum associated to ξ. Since

$$d(\phi_{g(t)}(x), \phi_t(p)) < \varepsilon, \quad \forall t > 0,$$

Lemma 6.6 implies that

$$d(\phi_t(p), \phi_{g(t)}\phi_{[-1,1]}(x)) \to 0 \quad \text{as } t \to \infty.$$

Hence, we can choose $k_1 \in \mathbb{N}$ such that $g(k_1T) > 1$ and

$$d(p, \phi_{g(k_1T)+\tau_1}(x)) = d(\phi_{k_1T}(p), \phi_{g(k_1T)+\tau_1}(x)) < \varepsilon$$

for some $\tau_1 \in [-1, 1]$. Similarly, since

$$d(\phi_t(p_\varepsilon), \phi_{g(t)}\phi_{[-1,1]}(x)) \to 0 \quad \text{as } t \to -\infty,$$

we can choose $k_2 \in \mathbb{N}$ such that $k_2S > 1$ and

$$d(p_\varepsilon, \phi_{g(-k_2S)+\tau_2}(x)) = d(\phi_{-k_2S}(p_\varepsilon), \phi_{g(-k_2S)+S}(x)) < \varepsilon$$

for some $\tau_2 \in [-1, 1]$. It follows that the sequence

$$\{(p_\varepsilon, S), (\phi_{g(-k_2S)+\tau_2}(x), g(k_1T) + \tau_1 - g(-k_2S) - \tau_2), (p, T)\}$$

is an $(\varepsilon, 1)$-chain from p_ε to p.

Similarly, we can construct an $(\varepsilon, 1)$-chain from p to p_ε. Hence, $p_\varepsilon \sim^\varepsilon p$. Since ε is arbitrary, it follows that $p \sim y$, and therefore $p \in \Omega_\lambda$. Consequently, we have

$$U_\lambda = U_\lambda \cap \overline{Per(\phi)} \subset \overline{U_\lambda \cap Per(\phi)} \subset \Omega_\lambda.$$

This shows that Ω_λ is open in $\Omega(\phi)$. Since $\Omega(\phi)$ is compact, it can be expressed by a finite disjoint union

$$\Omega(\phi) = \bigcup_{i=1}^{m} \Omega_i.$$

Next, we show that ϕ is topologically transitive on Ω_i for each $1 \leq i \leq m$. Let U, V be nonempty open subsets of Ω_i, and choose $r > 0$ such that $B(x, r) \cap \Omega_i \subset U$ and $B(y, r) \cap \Omega_i \subset V$ for some $x \in U$ and $y \in V$. By the shadowing of $\phi|_{\Omega(\phi)}$, take a constant $\gamma > 0$ such that every $(\gamma, 1)$-pseudo orbit of ϕ in $\Omega(\phi)$ is r-shadowed by a point in $\Omega(\phi)$. Since $x, y \in \Omega_i$, there exists a $(\gamma, 1)$-pseudo orbit $\{(w_i, t_i)\}_{i=0}^n$ in Ω_i such that $w_0 = x$ and $w_n = y$. We extend it to a $(\gamma, 1)$-pseudo orbit $\chi = \{(y_i, s_i)\}_{i\in\mathbb{Z}}$ in Ω_i by defining

$$(y_i, s_i) = \begin{cases} (\phi_i(x),\ 1), & i < 0, \\ (w_i,\ t_i), & 0 \leq i \leq n, \\ (\phi_{i-n}(y),\ 1), & i > n. \end{cases}$$

By the shadowing of ϕ, there exist $z \in \Omega(\phi)$ and $h \in \mathcal{H}$ such that

$$d(\phi_{h(t)}(z), \phi_{t-T_i}(y_i)) < r \text{ for all } T_i \leq t < T_{i+1} \text{ with } i \in \mathbb{Z},$$

where T_i is the ith sum associated to χ. It follows that $d(x, z) < r$ and $d(\phi_{h(T_n)}(z), y) < r$. Hence,

$$\phi_{h(T_n)}(B(x, r)) \cap B(y, r) \cap \Omega_i \subset \phi_{h(T_n)}(U) \cap V \neq \emptyset.$$

This shows that ϕ is topologically transitive on Ω_i. This completes the proof.

□

We note that if ϕ is expansive and shadowing on its nonwandering set $\Omega(\phi)$, then we cannot guarantee the existence of the spectral decomposition of ϕ as we see in Remark 6.2. For this, we first prove the following proposition.

Proposition 6.2 *Let ϕ be a flow on a compact metric space X which is expansive and shadowing on its nonwandering set $\Omega(\phi)$. If $\Omega(\phi)$ admits the spectral decomposition, then we have $\Omega(\phi) = \overline{Per(\phi)}$.*

Proof. Suppose the nonwandering set $\Omega(\phi)$ is decomposed as a finite union of disjoint closed invariant sets

$$\Omega(\phi) = \bigcup_{i=1}^{l} \Omega_i,$$

such that ϕ is topologically transitive on Ω_i for each $1 \leq i \leq l$. Choose a small constant $r > 0$ such that the r-neighborhoods of Ω_i $(i = 1, \ldots, l)$ are pairwise disjoint. Let $e > 0$ be an expansive constant of $\phi|_{\Omega(\phi)}$. For any $0 < \varepsilon < \min\{r, e/2\}$, take a constant $\delta > 0$ corresponding to ε by the shadowing of ϕ on the set $\Omega(\phi)$. Let $x \in \Omega(\phi)$. Then there exists an index $j \in \{1, \cdots, l\}$ such that $x \in \Omega_j$. Since ϕ is topologically transitive on Ω_j, we can construct a finite $(\delta, 1)$-pseudo orbit $\{(x_i, t_i)\}_{i=0}^{n-1}$ in Ω_j from x to itself. We extend it to a $(\delta, 1)$-pseudo orbit $\xi = \{(x_i, t_i)\}_{i \in \mathbb{Z}}$ by defining $x_i = x_{i+kn}$ and $t_i = t_{i+kn}$ for all $k \in \mathbb{Z}$ and $0 \leq i \leq n-1$. Since ϕ is shadowing on $\Omega(\phi)$, there exist $y \in X$ and $h_0 \in \mathcal{H}(\varepsilon)$ such that

$$d(\phi_{h_0(t)}(y), \phi_{t-T_i}(x_i)) \leq \varepsilon, \quad \forall t \in [T_i, T_{i+1}), \ i \in \mathbb{Z},$$

where T_i is the ith sum associated to ξ. For each $i \in \mathbb{N}$, define a homeomorphism $h_i : \mathbb{R} \to \mathbb{R}$ by

$$h_i(t) = h_0(t + T_i) - h_0(T_i), \quad \forall t \in \mathbb{R}.$$

We observe that $h_i \in \mathcal{H}(\varepsilon)$ for all $i \in \mathbb{N}$, and that the family $\{h_{in}\}_{i \in \mathbb{N}}$ is pointwise bounded and equicontinuous. By the Arzela–Ascoli theorem and the compactness of X, there exists a sequence $\{i_k\}_{k \in \mathbb{N}}$ such that

- $\phi_{h_0(T_{i_k n})}(y)$ converges to p as $k \to \infty$;
- $h_{i_k n}$ converges compactly to a homeomorphism h of $\mathbb{R}$ as $k \to \infty$.

Clearly, $h \in \mathcal{H}(\varepsilon)$. For any fixed $t \in \mathbb{R}$, write $t = T_{mn} + s$ for some $m \in \mathbb{Z}$ and $s \in [T_{k_0}, T_{k_0+1})$ for some $k_0 \in \{0, \cdots, n-1\}$. Then

$$\begin{aligned} & d(\phi_{h(t)}(p), \phi_{t-T_{mn+k_0}}(x_{mn+k_0})) \\ &= \lim_{k \to \infty} d(\phi_{h_0(T_{(m+i_k)n+k_0}+s)}, \phi_{t-T_{(m+i_k)n+k_0}}(x_{(m+i_k)n+k_0})) \\ &\leq \varepsilon. \end{aligned}$$

We note that $p \in \Omega_j$, since $d(x, p) = \lim_{k \to \infty} d(x_0, \phi_{h_0(T_{i_k n})}(y)) \leq \varepsilon < r$. Now, choose $k \in \mathbb{N}$ such that $h(kT_n) > 1$, and define a map $g \in \mathcal{H}$ by

$$g(t) = h(h^{-1}(t) + kT_n) - h(kT_n), \quad \forall t \in \mathbb{R}.$$

Letting $s = h^{-1}(t)$, we obtain

$$\begin{aligned}&d(\phi_t(p), \phi_{g(t)}(\phi_{h(kT_n)}(p)))\\&\leq d(\phi_{h(s)}(p), \phi_{s-T_i}(x_i)) + d(\phi_{s+kT_n-T_{i+kn}}(x_{i+kn}), \phi_{h(s+kT_n)}(p)) \leq e.\end{aligned}$$

Hence we get $\phi_{h(kT_n)}(p) \in \phi_{(-1,1)}(p)$, and thus p is a periodic point of ϕ. Since ε is arbitrary, we conclude that $x \in \overline{Per(\phi)}$. This completes the proof.

□

Remark 6.2 Remarkably, Dankner constructed a flow ϕ on a compact smooth manifold M with $\dim M \geq 4$ such that the nonwandering set $\Omega(\phi)$ is hyperbolic for ϕ, and $\Omega(\phi) \neq \overline{Per(\phi)}$ (see Theorem 6.2 in [17] for details). Since $\Omega(\phi)$ is hyperbolic for ϕ, the flow ϕ is expansive and shadowing on $\Omega(\phi)$. By Proposition 6.2, it follows that Dankner's flow ϕ does not admit the spectral decomposition even if it is both expansive and shadowing on its nonwandering set $\Omega(\phi)$.

Chapter 7
Measure Expansiveness for Flows

In this chapter, we study the notion of expansive measures for flows. In Sect. 7.1, we introduce the notion of expansive measures for flows, and prove some basic properties of expansive measures. In Sect. 7.2, we show that there is no measure expansive flow on a closed surface. Moreover, we give a characterization of measure expansive flows by using the notion of countable expansiveness.

7.1 Expansive Measures for Flows

Let ϕ be a flow on a compact metric space X, and $\mu \in \mathcal{M}(X)$. For any subset $B \subset X$ we write $\mu(B) = 0$ whenever $\mu(A) = 0$ for every Borel set $A \subset B$.

Definition 7.1 A Borel measure $\mu \in \mathcal{M}(X)$ is said to be *expansive* for ϕ (or ϕ is μ-expansive) if there is $\delta > 0$ such that $\mu(\Gamma_\delta^\phi(x)) = 0$ for all $x \in X$. The constant δ is called an *expansive constant* of μ.

We now provide some basic properties of expansive measures for flows. First, we recall some notions as follows.

Definition 7.2 A Borel measure $\mu \in \mathcal{M}(X)$ is said to be *ϕ-orbit vanishing* if $\mu(\phi_{\mathbb{R}}(x)) = 0$ for all $x \in X$. Denote by $\mathcal{M}_\phi^*(X)$ the collection of ϕ-orbit vanishing measures for ϕ.

Example 7.1 Let ϕ be a geometric expansive flow on a compact metric space X. Then every ϕ-orbit vanishing measure μ on X is expansive for ϕ. In fact, let μ be a ϕ-orbit vanishing measure and $\delta > 0$ be a geometric expansive constant of ϕ. Since ϕ is geometric expansive, we have

$$\mu(\Gamma_\delta^\phi(x)) \leq \mu(\phi_{\mathbb{R}}(x)) = 0, \quad \forall x \in X.$$

K. Lee et al., *Topological Dynamics from a Measure-Theoretic Viewpoint*, Springer Asia Pacific Mathematics Series 14, https://doi.org/10.1007/978-981-95-6390-6_7

Definition 7.3 Let X and Y be compact metric spaces, and let ϕ and ψ be flows on X and Y, respectively. We say that ϕ and ψ are *equivalent* if there exists a homeomorphism $h : X \to Y$ which sends the orbits of ϕ onto the orbits of ψ. In the case, h is called an *equivalence* between ϕ and ψ.

Observe that if $g : X \to Y$ is an equivalence between the flows ϕ and ψ on X and Y, respectively, then $g(Fix(\phi)) = Fix(\psi)$. Furthermore, if $\hat{g}(\psi)$ is the flow on X defined by

$$\hat{g}(\psi)(x, t) = g^{-1}(\psi(g(x), t)),$$

then for every $x \in X$ there is a homeomorphism $h_x \in \mathcal{H}$ satisfying

$$\hat{g}(\psi)_t(x) = \phi_{h_x(t)}(x), \quad \forall t \in \mathbb{R}. \tag{7.1}$$

Indeed, for $x \notin Fix(\phi)$, the existence of such an h_x was proved in [11], whereas for $x \in Fix(\phi)$ we can take $h_x(t) = t$. We shall use this in the following lemma.

Lemma 7.1 *Let ϕ and ψ be two flows on compact metric spaces X and Y respectively, and $g : X \to Y$ be an equivalence between the flows ϕ and ψ. Then for every $\delta > 0$ there is $\alpha > 0$ such that*

$$g^{-1}(\Gamma_\alpha^\psi(z)) \subset \Gamma_\delta^\phi(g^{-1}(z)), \quad \forall z \in Y.$$

Proof Let $\delta > 0$. Since g^{-1} is uniformly continuous, there is $\alpha > 0$ such that if $d(z, w) < \alpha$ for $z, w \in Y$, then $d(g^{-1}(z), g^{-1}(w)) \le \delta$.

Let $z, w \in Y$ be such that $w \in \Gamma_\alpha^\psi(z)$. Then there is $c \in \mathcal{C}$ such that

$$d(\psi_t(z), \psi_{c(t)}(w)) \le \alpha, \quad \forall t \in \mathbb{R}.$$

Then we get

$$d(g^{-1}\phi_t(z), g^{-1}\psi_{c(t)}(w)) \le \delta, \quad \forall t \in \mathbb{R},$$

and so

$$d(\hat{g}(\psi)_t(g^{-1}(z)), \hat{g}(\psi)_{c(t)}(g^{-1}(w))) \le \delta, \quad \forall t \in \mathbb{R},$$

Since g is equivalence between ϕ and ψ, by (7.1), we have

$$d(\phi_{h_{g^{-1}(z)}(t)}(g^{-1}(z)), \phi_{(h_{g^{-1}(w)} \circ c)(t)}(g^{-1}(w))) \le \delta, \quad \forall t \in \mathbb{R}.$$

Since $h_{g^{-1}(z)}$ is a homeomorphism we can replace t by $h^{-1}_{g^{-1}(z)}(t)$ in the expression above to obtain

$$d(\phi_t(g^{-1}(z)), \phi_{\hat{c}(t)}(g^{-1}(w))) \le \delta, \quad \forall t \in \mathbb{R},$$

where $\hat{c} = h_{g^{-1}(w)} \circ c \circ h^{-1}_{g^{-1}(z)}$. It is clear that $\hat{c} \in \mathcal{C}$, we conclude that $g^{-1}(w) \in \Gamma_\delta^\phi(g^{-1}(z))$, and so

$$g^{-1}(\Gamma_\alpha^\psi(z)) \subset \Gamma_\delta^\phi(g^{-1}(z)), \quad \forall z \in Y.$$

This completes the proof.

□

For a flow ϕ and $T \in \mathbb{R} \setminus \{0\}$, we observe that the time T-map $\phi_T : X \to X$ defined by $\phi_T(x) = \phi(T, x)$ is a homeomorphism on X.

With the above definitions, we now state several fundamental properties of expansive measures for flows.

Theorem 7.1 *Let ϕ and ψ be two flow on a compact metric space X, and $\mu \in \mathcal{M}(X)$. Then the following properties hold:*

(1) If μ is expansive for ϕ, then supp$(\mu) \cap Fix(\phi) = \emptyset$.
(2) If μ is expansive for ϕ, then ϕ is aperiodic with respect to μ, i.e., $\mu(Per(\phi)) = 0$.
(3) If h is an equivalence between ϕ and ψ, then μ is expansive for ϕ if and only if $h_*\mu$ is expansive for ψ.
(4) If μ is expansive for ϕ, then μ is ϕ-orbit vanishing.
(5) If μ is expansive for ϕ, then μ is expansive for ϕ_T for every $T \in \mathbb{R} \setminus \{0\}$.

Before proving Theorem 7.1, let us make some remark in light of the preceding results. First, statement (1) provides a measurable version of the well-known property that every fixed point of an expansive flow is isolated (see Lemma 6.1). In (2), we establish the aperiodicity property of expansive measures, which plays an important tool in the ergodic theory of flows. Furthermore, statement (3) shows that expansiveness of measures is invariant under flow equivalence. On the other hand, (4) demonstrates that every expansive measure of ϕ is ϕ-orbit vanishing. Finally, in (5), we describe the relationship between expansiveness of measures for flows and for time-T map ϕ_T for $T \in \mathbb{R} \setminus \{0\}$. It is worth noting that the converse of (5) does not hold, as illustrated by the following example.

Example 7.2 Let $(\mathcal{A}, \phi)$ be a geometric Lorenz attractor. It is well known that ϕ admits an ergodic Borel probability measure $\mu \in \mathcal{M}(\mathcal{A})$ with positive entropy (see, for example, [38]). Such a measure μ is known to be mixing (see [30]), and consequently, it is ergodic for every time-T maps ϕ_T with $T \in \mathbb{R} \setminus \{0\}$. By Theorem 3.4, we deduce that for each $T \in \mathbb{R} \setminus \{0\}$, the measure μ is expansive for time-T maps ϕ_T. However, since the support of μ contains a fixed point, Theorem 7.1 (1) implies that μ is not expansive for the flow ϕ.

We now construct a class of examples of expansive measures for flows by using suspension flows. For this, we introduce some notations. Let f be a homeomorphism on a compact metric space X, and let $\phi^{\tau,f}$ denote the suspension flow of f on the space $X^{\tau,f}$ under the height function τ. We define an operator $T^{\tau,f} : \mu \mapsto T^{\tau,f}(\mu)$ which associates to each Borel measure μ on X a Borel measure $T^{\tau,f}(\mu)$ of $X^{\tau,f}$ defined implicitly by

$$\int_{X^{\tau,f}} h(y) dT^{\tau,f}(\mu)(y) = \int_X \frac{1}{\tau(x)} \int_0^{\tau(x)} h(\phi_t^{\tau,f}(x)) dt \, d\mu(x) \tag{7.2}$$

for every continuous map $h : X^{\tau,f} \to \mathbb{R}$. The measure $T^{\tau,f}(\mu)$ is called a *suspending measure*. An important property of the operator $T^{\tau,f}$ is that $\mu \in \mathcal{M}(X)$ is a probability (respectively, nonatomic, invariant) measure if and only if $T^{\tau,f}$ is a probability (respectively, orbit vanishing, invariant) measure on the suspension space $X^{\tau,f}$. For simplicity, when the height function is $\tau = 1$, we write T^f instead of $T^{\tau,f}$.

Theorem 7.2 *Let f be a homeomorphism on a compact metric space X, and ϕ be the suspension flow $\phi^{\tau,f}$ of f under the height function τ on $X^{\tau,f}$. Then a Borel measure $\mu \in \mathcal{M}(X)$ is expansive for f if and only if $T^{\tau,f}(\mu) \in \mathcal{M}(X^{\tau,f})$ is expansive for $\phi^{\tau,f}$.*

By using Theorem 7.2, we show that there exists an expansive measure for a nonexpansive flow ϕ.

Example 7.3 Let f be a 2-expansive homeomorphism on a compact metric space X as in Example 5.1, and let ϕ^f denote the suspension flow of f under the constant height function $\tau = 1$. Let μ be a nonatomic measure supported on X. Then the corresponding suspending measure $T^f(\mu)$ is a ϕ^f-orbit vanishing measure on X^f. Let $\delta > 0$ be a 2-expansive constant of f. For each $x \in X$, there exist $x_1, x_2 \in X$ such that $\Gamma_\delta^f(x) \subset \{x_1, x_2\}$. It follows that for any $\overline{x} = (x, s) \in X^f$, $\Gamma_\delta^{\phi^f}(\overline{x}) \subset \phi_{\mathbb{R}}(x_1, s) \cup \phi_{\mathbb{R}}(x_2, s)$. Then

$$T^f(\mu)\Gamma_\delta^{\phi^f}(x) \leq T^f(\mu)\phi_{\mathbb{R}}(x_1, s) + T^f(\mu)\phi_{\mathbb{R}}(x_2, s) = 0.$$

Hence, $T^f(\mu)$ is expansive for the suspension flow ϕ^f. On the other hand, since f is not expansive, Theorem 6.3 implies that ϕ^f is not expansive.

Lemma 7.2 *Let ϕ be a flow on a compact metric space X, and $\mu \in \mathcal{M}(X)$. If μ is expansive for ϕ, then μ is ϕ-orbit vanishing.*

Proof Let μ be an expansive measure of ϕ with expansive constant $\delta > 0$. We first show that there exists a constant $\alpha > 0$ such that $d(\phi_s(x), x) < \delta$ for all $x \in X$ and $s \in (-\alpha, \alpha)$. Suppose, on the contrary, that for every $n \in \mathbb{N}$, there exist $x_n \in X$ and $s_n \in (-\frac{1}{n}, \frac{1}{n})$ such that $d(\phi_{s_n}(x), x) \geq \delta$ for all $n \in \mathbb{N}$. Since X is compact, we may assume that $x_n \to y$ for some $y \in X$. Then, by the continuity of ϕ,

$$0 = d(y, y) = \lim_{n\to\infty} d(\phi_{s_n}(x_n), x_n) \geq \delta,$$

which is a contradiction. Hence, there exists $\alpha > 0$ such that $\phi_{(-\alpha,\alpha)}(x) \subset \Gamma_\delta^\phi(x)$ for all $x \in X$. Since μ is expansive, this implies $\mu(\phi_{(-\alpha,\alpha)}(x)) = 0$ for all $x \in X$.

For a given $x \in X$, there are two possibilities: either $\phi_{\mathbb{R}}(x)$ is a closed subset of X or the induced map $\phi^x : \mathbb{R} \to \phi_{\mathbb{R}}(x)$ defined by $\phi^x(t) = \phi(x, t)$ is a homeomorphism onto $\phi_{\mathbb{R}}(x)$. In either case, we can find sequences $\{x_k\}_{k\in\mathbb{N}} \subset X$ and $\{\delta_k\}_{k\in\mathbb{N}} \subset (0, \alpha)$ such that the collection $\{\phi_{(-\delta_k,\delta_k)}(x_k) : k \in \mathbb{N}\}$ forms a cover of $\phi_{\mathbb{R}}(x)$. Therefore,

$$\mu(\phi_{\mathbb{R}}(x)) \leq \sum_{k\in\mathbb{N}} \mu(\phi_{(-\delta_k,\delta_k)}(x_k)) = 0.$$

Since $x \in X$ is arbitrary, we conclude that $\mu(\phi_{\mathbb{R}}(x)) = 0$, for all $x \in X$. This completes the proof.

□

Another property of the expansive measures is given below.

Lemma 7.3 *Let ϕ be a flow on a compact metric space X. For any expansive measure μ for ϕ,* supp(μ) *does not contain a fixed point of ϕ.*

Proof Let μ be an expansive measure of the flow ϕ with expansive constant $\delta > 0$. If $p \in Fix(\phi)$, then it is clear that $B(p, \delta) \subset \Gamma_\delta^\phi(p)$. Since μ is expansive for ϕ, we have $\mu(B(p, \delta)) \leq \mu(\Gamma_\delta^\phi(x)) = 0$. Hence, $p \notin \text{supp}(\mu)$.

□

By combining the proofs of Theorems 3 and 5 in [11], we obtain the following lemma. In its proof, we denote by $[r]$ the integer part of a real number $r \in \mathbb{R}$.

Lemma 7.4 *Let ϕ be a flow on a compact metric space X. Then, for every $\delta > 0$ and $t > 0$, there exist constants $0 < \alpha_0 < \min\{\delta, 2t\}$ and $0 < \beta_0 < \min\{\delta, 2t\}$ such that if $x, y \in X$ satisfy $\phi_a(x) = x$, $\phi_b(y) = y$ for some $a, b \in \left[t - \frac{\alpha}{2}, t + \frac{\alpha}{2}\right]$ and $d(x, y) \leq \beta$ where $0 < \alpha < \alpha_0$ and $0 < \beta < \beta_0$, then $y \in \Gamma_\delta^\phi(x)$.*

Proof Let $\delta > 0$ and $t > 0$. Choose $0 < \beta_0 < \min\{\delta, 2t\}$ such that if $z, w \in X$ satisfy $d(z, w) \leq \beta$ for some $0 < \beta < \beta_0$, then

$$d(\phi_s(z), \phi_s(w)) \leq \frac{\delta}{2}, \quad \forall 0 \leq s \leq t + 1. \tag{7.3}$$

Next, choose $0 < \alpha_0 < \min\{\delta, 2t\}$ such that for every $0 < \alpha < \alpha_0$,

$$\max\left\{\alpha, 2\left(\sup_{(z,u)\in X\times[0,\alpha]} d(\phi_u(z), z)\right)\right\} < \min\left\{\frac{\delta}{2}, 1\right\}. \tag{7.4}$$

Now fix $x, y \in X$ satisfying $\phi_a(x) = x, \phi_b(y) = y$ for some $a, b \in \left[t - \frac{\alpha}{2}, t + \frac{\alpha}{2}\right]$ and such that $d(x, y) \leq \beta$ with $0 < \alpha < \alpha_0$ and $0 < \beta < \beta_0$.

Let $m = \left[\frac{t - \frac{\alpha}{2}}{\alpha}\right] + 1$ and define

$$t_{pm+q} = pa + q\alpha \text{ and } u_{pm+q} = pb + q\alpha, \quad \forall (p, q) \in \mathbb{Z} \times \{0, \ldots, m-1\}.$$

A direct computation gives $t_{pm+q+1} - t_{pm+q} = \alpha$ (for $(p, q) \in \mathbb{Z} \times \{0, \ldots, m-2\}$), $t_{(p+1)m} - t_{pm+m-1} \geq 0$ (for $p \in \mathbb{Z}$), and analogously replacing t by u. Moreover,

$$|t_{pm+q+1} - t_{pm+q}| \leq \alpha \quad \text{and} \quad |u_{pm+q+1} - u_{pm+q}| \leq \alpha \tag{7.5}$$

for every $(p, q) \in \mathbb{Z} \times \{0, \ldots, m-1\}$. Also note that

$$d(\phi_{t_{pm+q}}(x), \phi_{u_{pm+q}}(y)) = d(\phi_{q\alpha}(x), \phi_{q\alpha}(y)).$$

Since

$$0 \leq q\alpha < m\alpha = \left[\frac{t - \frac{\alpha}{2}}{\alpha}\right]\alpha + \alpha = t - \frac{\alpha}{2} - r + \alpha$$

for some $0 \leq r < \alpha$, we have

$$0 \leq q\alpha < t + \frac{\alpha}{2} \leq t + 1, \quad \forall q = 1, \ldots, m - 1.$$

Because $d(x, y) \leq \beta$, we may substitute $s = q\alpha$ in (7.3) to obtain

$$d(\phi_{q\alpha}(x), \phi_{q\alpha}(y)) \leq \frac{\delta}{2},$$

and so

$$d(\phi_{t_{pm+q}}(x), \phi_{u_{pm+q}}(y)) \leq \frac{\delta}{2}, \quad \forall (p, q) \in \mathbb{Z} \times \{0, \ldots, m - 1\}. \tag{7.6}$$

Now define $h : \mathbb{R} \to \mathbb{R}$ by setting $h(t_{pm+q}) = u_{pm+q}$ and extending h linearly on each interval $[t_{pm+q}, t_{pm+q+1}]$, for all $(p, q) \in \mathbb{Z} \times \{0, \ldots, m - 1\}$. Clearly $h \in \mathcal{C}$. Moreover, for each $s \in \mathbb{R}$, there exists a unique pair $(p, q) \in \mathbb{Z} \times \{0, \ldots, m - 1\}$ such that $s \in [t_{pm+q}, t_{pm+q+1})$. Then, from (7.4) to (7.6), we have

$$\begin{aligned} d(\phi_s(x), \phi_{h(s)}) &\leq d(\phi_s(x), \phi_{t_{pm+q}}(x)) + d(\phi_{t_{pm+q}}(x), \phi_{u_{pm+q}}(y)) \\ &\quad + d(\phi_{u_{pm+q}}(y), \phi_{h(s)}(y)) \\ &\leq 2\left(\sup_{(z,u)\in X\times[0,\alpha]} d(\phi_u(z), z)\right) + d(\phi_{t_{pm+q}}(x), \phi_{u_{pm+q}}(y)) \\ &\leq \frac{\delta}{2} + \frac{\delta}{2} = \delta. \end{aligned}$$

Hence, there exists $h \in \mathcal{C}$ satisfying

$$d(\phi_s(x), \phi_{h(s)}(y)) < \delta, \quad \forall s \in \mathbb{R}.$$

Therefore, $y \in \Gamma_\delta^\phi(x)$.

□

We also require a lemma concerning the localization of the periods of the periodic points. More precisely, for a flow ϕ and a point $x \in Per(\phi)$, we define

$$t_x = \min\{t > 0 : \phi_t(x) = x\},$$

that is, t_x denotes the period of x. For a given interval $I \subset \mathbb{R}$, we define

$$Per_I(\phi) = \{x \in Per(\phi) : t_x \in I\}.$$

For simplicity, when $N > 0$, we write $Per_N(\phi)$ instead of $Per_{[0,N]}(\phi)$.

Corollary 7.1 *Let ϕ be a flow on a compact metric space X. Then, for every $\delta > 0$, $N \in \mathbb{N}$ and $x \in Per(\phi) \setminus Fix(\phi)$, there are $0 < \alpha_* < \min\{\delta, 2t_x\}$ and $0 < \beta_* < \min\{\delta, 2t_x\}$ such that*

$$B[x, \beta] \cap \left(\bigcup_{k=1}^{N} Per_{[kt_x - \frac{\alpha}{2}, kt_x + \frac{\alpha}{2}]}(\phi) \right) \subset \Gamma_\delta^\phi(x),$$

where $0 < \alpha < \alpha_*$ and $0 < \beta < \beta_*$.

Proof Let $\delta > 0$, $N \in \mathbb{N}$, and $x \in Per(\phi) \setminus Fix(\phi)$. For each $k = 1, \dots, N$, apply Lemma 7.4 with $t = kt_x$ to obtain the corresponding constants $0 < \alpha_k < \min\{\delta, 2kt_x\}$ and $0 < \beta_k < \min\{\delta, 2kt_x\}$. Define

$$\alpha_* = \min_{1 \le k \le N} \alpha_k \quad \text{and} \quad \beta_* = \min_{1 \le k \le N} \beta_k.$$

Then clearly $0 < \alpha_* < \min\{\delta, 2t_x\}$ and $0 < \beta_* < \min\{\delta, 2t_x\}$. Fix any $0 < \alpha < \alpha_*$ and $0 < \beta < \beta_*$. Take $y \in B[x, \beta] \cap Per_{[kt_x - \frac{\alpha}{2}, kt_x + \frac{\alpha}{2}]}(\phi)$ for some $k = 1, \dots, N$. Setting $a = kt_x$ and $b = t_y$, we have $\phi_a(x) = x$ and $\phi_b(y) = y$ with $a, b \in [kt_x - \frac{\alpha}{2}, kt_x + \frac{\alpha}{2}]$. Since $0 < \alpha < \alpha_* \le \alpha_k$ and $0 < \beta < \beta_* \le \beta_k$, we can apply Lemma 7.4 with $t = kt_x$ to conclude that $y \in \Gamma_\delta^\phi(x)$. This completes the proof. □

Another localization lemma for periodic points is given below.

Lemma 7.5 *Let ϕ be a flow on a compact metric space X. Then, for every $T > 0$, $x \in Per_T(\phi) \setminus Fix(\phi)$, and $0 < \alpha < t_x$, there exists $\beta_0 > 0$ such that*

$$B[x, \beta] \cap Per_T(\phi) \subset \bigcup_{k=1}^{\left[\frac{T}{t_x}\right]} Per_{[kt_x - \frac{\alpha}{2}, kt_x + \frac{\alpha}{2}]}(\phi), \quad \forall 0 < \beta < \beta_0.$$

Proof Suppose not. Then there exists $T > 0$, $x \in Per_T(\phi) \setminus Fix(\phi)$, and $0 < \alpha < t_x$ such that

$$x_n \notin \bigcup_{k=1}^{\left[\frac{T}{t_x}\right]} Per_{[kt_x - \alpha, kt_x + \alpha]}(\phi), \tag{7.7}$$

for some sequence $x_n \in Per_T(\phi)$ with $x_n \to x$. It is clear that $t_{x_n} \in [0, T]$ for all $n \in \mathbb{N}$. Then we can assume that $t_{x_n} \to t_\infty$ for some $t_\infty \in [0, T]$. Since $x \notin Fix(\phi)$ and $x_n \to x$, it follows that $t_\infty > 0$. Moreover, since $\phi_{t_{x_n}}(x_n) = x_n \to x$ and $\phi_{t_{x_n}}(x_n) \to \phi_{t_\infty}(x)$, we conclude that $\phi_{t_\infty}(x) = x$, that is, $x \in Per(\phi)$ with $t_\infty = kt_x$ for some

$1 \le k \le \left[\frac{T}{t_x}\right]$. Hence, for all sufficiently large n, we have $t_{x_n} \in [kt_x - \alpha, kt_x + \alpha]$, which means $x_n \in Per_{[kt_x-\alpha, kt_x+\alpha]}(\phi)$, contradicting (7.7). Therefore, the constant β_0 exists.

□

Lemma 7.6 *Let ϕ be a flow on a compact metric space X, and let $\mu \in \mathcal{M}(X)$. If μ is expansive for ϕ, then $\mu(Per_T(\phi)) = 0$ for every $T \ge 0$.*

Proof We first observe that $Per_T(\phi)$ is compact for all $T \ge 0$. Suppose. for contradiction, that there exists $T \ge 0$ such that $\mu(Per_T(\phi)) > 0$. Since $Per_T(\phi)$ is compact, there exists $x_* \in Per_T(\phi)$ and $\beta_- > 0$ such that

$$\mu(B[x_*, \beta] \cap Per_T(\phi)) > 0, \quad \forall 0 < \beta < \beta_-. \tag{7.8}$$

Hence $x_* \in \operatorname{supp}(\mu)$, and by Lemma 7.3, it follows that $x_* \notin Fix(\phi)$. In particular, $T > 0$.

Let $\delta > 0$ be an expansive constant of μ, and take $N = \left[\frac{T}{t_{x_*}}\right]$. Let α_* and β_* be the constants associated with $x = x_*$ as in Corollary 7.1. Fix $0 < \alpha < \alpha_*$, which together with $x = x_*$ and T, determines a corresponding constant β_0 as in Lemma 7.5. Now choose

$$0 < \beta < \min\{\beta_0, \beta_-, \beta_*\}.$$

On the one hand, by Lemma 7.5, we have

$$B[x_*, \beta] \cap Per_T(\phi) \subset \bigcup_{k=1}^{\left[\frac{T}{t_x}\right]} Per_{[kt_{x_*}-\frac{\alpha}{2}, kt_{x_*}+\frac{\alpha}{2}]}(\phi),$$

and hence

$$B[x_*, \beta] \cap Per_T(\phi) \subset B[x_*, \beta] \cap \left(\bigcup_{k=1}^{\left[\frac{T}{t_x}\right]} Per_{[kt_{x_*}-\frac{\alpha}{2}, kt_{x_*}+\frac{\alpha}{2}]}(\phi) \right).$$

On the other hand, since $0 < \alpha < \alpha_*$, we may take $N = \left[\frac{T}{t_{x_*}}\right]$ in Corollary 7.1 to obtain

$$B[x_*, \beta] \cap \left(\bigcup_{k=1}^{\left[\frac{T}{t_x}\right]} Per_{[kt_{x_*}-\frac{\alpha}{2}, kt_{x_*}+\frac{\alpha}{2}]}(\phi) \right) \subset \Gamma_\delta^\phi(x_*).$$

Combining these inclusions gives

$$\mu(B[x_*, \beta] \cap Per_T(\phi)) \le \mu\left(B[x_*, \beta] \cap \left(\bigcup_{k=1}^{\left[\frac{T}{t_{x*}}\right]} Per_{[kt_{x_*} - \frac{\alpha}{2}, kt_{x_*} + \frac{\alpha}{2}]}(\phi)\right)\right)$$
$$\le \mu(\Gamma_\delta^\phi(x_*)) = 0.$$

This contradicts (7.8), since $0 < \beta < \beta_-$ is arbitrary. Therefore, $\mu(Per_T(\phi)) = 0$ for all $T \ge 0$.

□

Lemma 7.7 *Let ϕ be a flow on a compact metric space X. Then, for every $\delta > 0$ and $T \in \mathbb{R} \setminus \{0\}$, there exists $\alpha > 0$ such that*

$$\hat{\Gamma}_\alpha^{\phi_T}(x) \subset \Gamma_\delta^\phi(x), \quad \forall x \in X.$$

Proof Let $\varepsilon > 0$ be such that if $d(x, y) < \varepsilon$ for $x, y \in X$, then

$$d(\phi_s(x), \phi_s(y)) \le \delta, \quad \forall s \in [0, T]. \tag{7.9}$$

Let $x, y \in X$ satisfy $y \in \hat{\Gamma}_\alpha^{\phi_{nT}}(x)$, i.e.,

$$d(\phi_{nT}(x), \phi_{nT}(y)) \le \varepsilon, \quad \forall n \in \mathbb{Z}. \tag{7.10}$$

For $t \in \mathbb{R}$, there exists $n \in \mathbb{Z}$ such that $nT \le t < (n+1)T$. Set $s = t - nT \in [0, T)$, and define $z = \phi_{nT}(x)$ and $w = \phi_{nT}(y)$. By (7.9), $d(z, w) \le \varepsilon$, and thus by (7.10),

$$d(\phi_t(x), \phi_t(y)) = d(\phi_s(z), \phi_s(w)) \le \delta.$$

Hence, $y \in \Gamma_\delta^\phi(x)$.

□

Proof of Theorem 7.1

(1) The proof follows directly from Lemma 7.3.
(2) Let μ be an expansive measure of ϕ. We note that $Per(\phi) = \bigcup_{N \in \mathbb{N}} Per_N(\phi)$. By Lemma 7.6, we obtain that

$$\mu(Per(\phi)) = \sum_{N \in \mathbb{N}} \mu(Per_N(\phi)) = 0.$$

This completes the proof.
(3) Let μ an expansive measure of ϕ with an expansive constant δ, and α be as in Lemma 7.1 with respect to δ. Let $z \in Y$ and a Borel set $B \subset \Gamma_\alpha^\psi(z)$. By Lemma 7.1, we have $f^{-1}(B) \subset \Gamma_\delta^\phi(f^{-1}(z))$, and so

$$f_*\mu(B) = \mu(f^{-1}(B)) \le \mu(\Gamma_\delta^\phi(f^{-1}(z))) = 0.$$

It follows that $f_*\mu(\Gamma_\alpha^\psi(z)) = 0$, $\forall z \in Y$. Hence $f_*\mu$ is expansive for ψ with expansive constant α.

(4) The proof follows directly from Lemma 7.2.

(5) Let μ be an expansive measure of ϕ with expansive constant δ, and $T \in \mathbb{R} \setminus \{0\}$. Take $\alpha > 0$ as in Lemma 7.7 with respect to such δ and T. Then we have

$$\mu(\hat{\Gamma}_\alpha^{\phi_T}(x)) = 0, \quad \forall x \in X.$$

Hence μ is expansive for ϕ_T.

□

We are now at the position to prove Theorem 7.2. For this, we need the following remark.

Remark 7.1 Let (X, d) be a compact metric space. We define a distance d' on X by

$$d'(x, y) = \min\{d(x, y), d(f(x), f(y))\}, \quad \forall x, y \in X.$$

We observe that a measure $\mu \in \mathcal{M}(X)$ is expansive for a homeomorphism f if and only if there is $\delta > 0$ such that $\mu(\hat{\Gamma}_\delta^f(x)) = 0$ for all $x \in X$, where

$$\hat{\Gamma}_\delta^f(x) = \{y \in X : d(f^n(x), f^n(y)) \leq \delta, \quad \forall n \in \mathbb{Z}\}.$$

Proof of Theorem 7.2 We first observe that there exists a natural equivalence $\lambda : X^{\tau,f} \to X^f$ between $\phi^{\tau,f}$ and ϕ^f, defined by

$$\lambda(x, t) = \left(x, \frac{t}{\tau(x)}\right).$$

It is easy to verify that

$$\lambda_*\mu^{\tau,f} = \mu^{1,f}.$$

Hence, by Theorem 7.1(3), it suffices to consider the case $\tau = 1$.

Assume that $T^f(\mu)$ is expansive for the suspension flow ϕ^f. By Theorem 7.1(5), it follows that $T^f(\mu)$ is also expansive for the time-one map ϕ_1^f. Denote by Id the identity map on $\left[0, \frac{1}{2}\right]$. For any $x \in X$ and $0 \leq s \leq \frac{1}{2}$, we have

$$\begin{aligned} \phi_1^f(x, s) &= (x, s + 1) = (f^{[s+1]}(x), s + 1 - [s + 1]) \\ &= (f(x), s) = (f \times Id)(x, s). \end{aligned}$$

Therefore, $T^f(\mu)$ is also expansive for

$$f \times Id : X \times \left[0, \frac{1}{2}\right] \to X \times \left[0, \frac{1}{2}\right].$$

Since $T^f(\mu)$ restricted to $X \times \left[0, \frac{1}{2}\right]$ coincides with the product measure $\mu \times m$, where m denotes the Lebesgue measure on $[0, \frac{1}{2}]$, it follows that $\mu \times m$ is expansive for $f \times Id$. Consequently, μ is expansive for f.

Conversely, suppose that μ is expansive for f. Let d' be the distance on X as in Remark 7.1. Then there exists a constant $0 < \delta < \frac{1}{4}$ such that

$$\mu(\hat{\Gamma}_\delta^f(x)) = 0, \quad \forall x \in X.$$

Now let $y = (x, t) \in X^f$. By Lemma 6.2, we have $\Gamma_\delta^\phi(y) \subset \hat{\Gamma}_\delta^f(x) \times [0, 1]$. Hence, for any Borel set $B \subset \Gamma_\delta^\phi(y)$, $B \subset \hat{\Gamma}_\delta^f(x) \times [0, 1]$ and therefore $\chi_B \leq \chi_{\hat{\Gamma}_\delta^f(x) \times [0,1]}$ where χ denotes the characteristic function. Setting $\tau = 1$ in (7.2), we obtain

$$\begin{aligned} T^f(\mu)(B) &= \int_X \int_0^1 \chi_B(\phi_t^f(z))dtd\mu(z) \\ &\leq \int_X \int_0^1 \chi_{\hat{\Gamma}_\delta^f(x) \times [0,1]}(\phi_t^f(z))dtd\mu(z) \\ &= \int_X \chi_{\hat{\Gamma}_\delta^f(x)}(z)d\mu(z) = \mu(\hat{\Gamma}_\delta^f(x)) = 0. \end{aligned}$$

Thus $T^f(\mu)(\Gamma_\delta^\phi(y)) = 0$. Since $y \in X^f$ is arbitrary, we conclude that $T^f(\mu)$ is expansive for ϕ^f with expansive constant δ.

□

7.2 Measure Expansive Flows

Let ϕ be a flow on a compact metric space X. By Theorem 7.1(4), we observe that if ϕ is expansive then every ϕ-orbit vanishing measure $\mu \in \mathcal{M}(X)$ is expansive for ϕ. However, the converse is not true in general as we can see in the following example.

Example 7.4 In this example, we show that there exists a flow ϕ on a compact metric space such that ϕ is not expansive, and every ϕ-orbit vanishing measure is expansive for ϕ. Let f be a homeomorphism on a compact metric space as in Example 6.1. We observe that f is not expansive, since for every $\delta > 0$, there exist points $x_n = (0, a_{n0})$ and $y_n = (0, a_{n1})$ in X such that $x_n \notin \Gamma_\delta^f(y_n)$. Let ϕ^f denote the suspension of f under the constant height function $\tau = 1$. By Theorem 6.3, it follows that ϕ^f is not expansive.

On the other hand, as in Example 6.1, we have

$$\Gamma_{1/2}^{\phi^f}(\overline{x}) \subset \phi_{\mathbb{R}}(\overline{x}), \quad \forall \overline{x} \in X^f.$$

Hence, for every $\mu \in \mathcal{M}_{\phi^f}^*(X^f)$,

$$\mu(\Gamma^{\phi^f}_{1/2}(\overline{x})) \leq \mu(\phi_{\mathbb{R}}(\overline{x})) = 0, \quad \forall \overline{x} \in X^f.$$

Therefore, every ϕ^f-orbit vanishing measure μ is expansive for ϕ^f.

In this section, we investigate the class of flows for which every orbit vanishing measure is expansive. More precisely, we first show that there is no measure expansive flow on closed surfaces. Furthermore, we provide a characterization of measure expansive flows by using the notion of countable expansiveness.

Definition 7.4 We say that a flow ϕ on a compact metric space X is *measure expansive* if every ϕ-orbit vanishing Borel measure on X is expansive for ϕ.

We observe that every expansive flow on a compact metric space is measure expansive. However, the converse is not true as we can see in Example 7.4.

In the following proposition, we show that if ϕ is measure expansive, then there exists a uniform expansive constant for every ϕ-orbit vanishing measures.

Proposition 7.1 *Let ϕ be a flow on a compact metric space X. Then a flow ϕ is measure expansive if and only if there is a constant $\delta > 0$ such that $\mu(\Gamma^{\phi}_{\delta}(x)) = 0$ for all $\mu \in \mathcal{M}^*_{\phi}(X)$ and $x \in X$.*

Proof Suppose that ϕ is measure expansive. Assume, on the contrary, that for each $n \in \mathbb{N}$, there exist $x_n \in X$, $\mu_n \in \mathcal{M}^*_{\phi}(X)$ and a Borel set $B_n \subset \Gamma^{\phi}_{\delta}(x_n)$ such that $\mu(B_n) > 0$. Define a Borel measure μ on X by

$$\mu(A) = \sum_{n=1}^{\infty} \frac{\mu_n(A)}{2^n}$$

for any Borel subset A of X. It is clear that μ is ϕ-orbit vanishing. Since ϕ is measure expansive, μ is expansive for ϕ. Let $\delta > 0$ be an expansive constant of μ, and choose $N \in \mathbb{N}$ such that $\frac{1}{N} < \delta$. Then we have

$$\mu(B_N) \geq \frac{\mu_N(B_N)}{2^N} > 0,$$

which is a contradiction. This completes the proof. □

It is well-known that there are no expansive flows of closed surfaces [20]. In the following theorem, we extend the result to measure expansive flows. The proof is based on [13].

Theorem 7.3 *There are no measure expansive flows on closed surfaces.*

Proof Suppose, on the contrary, that there exists a measure expansive flow ϕ on a closed surface. By Gutierrez's result [19] and Theorem 7.1(3), we can assume that ϕ is of class C^1. It follows that every orbit of ϕ is a submanifolds of nontrivial codimension. Since all such submanifolds have zero Lebesgue measure, we deduce that ϕ admits at least one Borel measure that both has full support and vanishes along the orbits. By Theorem 7.1(1), this implies that $Fix(\phi) = \emptyset$. Consequently, the underlying surface must be either a torus or a Klein bottle. Passing to a double covering if necessary, we can assume without loss of generality that the surface is the torus T^2.

Since ϕ is nontrivial, we can assume that it has nontrivial recurrence, that is, there exists a nonperiodic, nonfixed point x_0 such that $x_0 \in \omega(x_0, \phi)$. As x_0 is not periodic, there exists a transversal circle X to ϕ containing x_0 (see Proposition 5.1.2 in [21]). Associated with this transversal, we define the return time function $\tau : \mathrm{Dom}(\tau) \subset X \to (0, \infty)$, and the corresponding first return map $f : \mathrm{Dom}(f) \subset X \to X$, $f(x) = \phi_{\tau(x)}(x)$, where $\mathrm{Dom}(\cdot)$ denotes the domain of definition. Note that X cannot separate T^2, since ϕ has no fixed points.

Since $x_0 \in X$ is recurrent, we have $\mathrm{Dom}(\tau) \neq \emptyset$. If there existed a point $x_1 \in X \setminus \mathrm{Dom}(\tau)$, then, using the fact that X does not separate T^2 and applying the Poincaré–Bendixon Theorem [31], the positive orbit of x_1 would spiral toward a periodic orbit parallel to X. In that case, $\mathrm{Dom}(\tau) = \emptyset$, a contradiction. Hence, $\mathrm{Dom}(\tau) = X$, and therefore $\mathrm{Dom}(f) = X$. Consequently, X intersects every orbit of ϕ (see, for instance, p. 146 in [31]). Thus, there exists an equivalence R between ϕ and the suspension flow $\phi^{tau,f}$ over f with the height function τ.

Let μ be a nonatomic Borel measure on X, and let $T^{\tau,f}(\mu)$ denote its suspension measure. Then $R_*^{-1}(T^{\tau,f}(\mu))$ is ϕ-orbit vanishing. Since ϕ is measure expansive, it follows that $R_*^{-1}(T^{\tau,f}(\mu))$ is expansive for ϕ. Because R is an equivalence between ϕ and $\phi^{\tau,f}$, Theorem 7.1(3) implies that $T^{\tau,f}(\mu) = R_*(R_*^{-1}(T^{\tau,f}(\mu)))$ is expansive for $\phi^{\tau,f}$. Applying Theorem 7.2, we conclude that μ is expansive for f. Hence, every nonatomic Borel measure on X is expansive for f. However, by Example 3.2, there exists no measure expansive homeomorphism on the circle. This contradiction completes the proof.

□

In Chap. 3, we gave a characterization of measure expansive homeomorphisms by using the notion of countable expansiveness. More precisely, we showed that a homeomorphism f on a compact metric space X is measure expansive if and only if it is countably expansive (see Theorem 3.5). Here we introduce the notion of N-expansiveness and countable expansiveness for flows.

Definition 7.5 A flow ϕ on a compact metric space X is said to be *N-expansive*($N \in \mathbb{N}$) if there is $\delta > 0$ such that for any $x \in X$, $\Gamma_\delta^\phi(x) \subset \bigcup_{i=1}^N \phi_{\mathbb{R}}(x_i)$ for some $\{x_i\}_{i=1}^N \subset X$. Moreover, a flow ϕ is said to be *countably expansive* if there is $\delta > 0$ such that for any $x \in X$, $\Gamma_\delta^\phi(x) \subset \bigcup_{i=1}^\infty \phi_{\mathbb{R}}(x_i)$ for some $\{x_i\}_{i=1}^\infty \subset X$.

In the rest of this section, we also characterize the measure expansive flows by using the notion of countable expansiveness.

Theorem 7.4 *Let ϕ be a flow on a compact metric space X. Then ϕ is measure expansive if and only if it is countably expansive.*

To prove Theorem 7.4, we first introduce some notations and preliminary lemmas. For any $x \in X$ and $\delta > 0$, we recall a variant of the dynamic ball introduced in [22], defined by

$$\tilde{\Gamma}_\delta^\phi(x) = \bigcap_{r>0} \bigcap_{\gamma>\delta} \bigcup_{c\in\mathcal{C}} \bigcap_{|t|\le r} \phi_{-c(t)}(B[\phi_t(x), \gamma]).$$

In the following lemmas, we show that if a flow ϕ has no fixed points, then for every $x \in X$ and $\delta > 0$, there exists $\varepsilon > 0$ such that $\tilde{\Gamma}_\varepsilon^\phi(x)$ is compact and $\Gamma_\varepsilon^\phi(x) \subset \tilde{\Gamma}_\varepsilon^\phi(x) \subset \Gamma_\delta^\phi(x)$. This result is based on the corresponding arguments in [22].

Lemma 7.8 *Let ϕ be a flow on a compact metric space X without fixed points. Then for every $\delta > 0$, there is $\varepsilon > 0$ such that*

$$\Gamma_\varepsilon^\phi(x) \subset \tilde{\Gamma}_\varepsilon^\phi(x) \subset \Gamma_\delta^\phi(x), \quad \forall x \in X.$$

Proof Let $\delta > 0$, and take $T_0 > 0$ as in Lemma 6.4. By the continuity of ϕ, there exists $0 < \lambda < T_0$ such that

$$d(\phi_t(x), x) < \frac{\delta}{2}, \quad \forall x \in X \text{ and } |t| < \lambda. \tag{7.11}$$

Moreover, by Lemma 6.4, there exists a constant $\gamma > 0$ such that

$$d(\phi_\lambda(x), x) > \gamma, \quad \forall x \in X. \tag{7.12}$$

Choose $m \in \mathbb{N}$ such that $\delta < \gamma m$, and set $\varepsilon = \frac{\delta}{3m}$. Let $x, z \in X$ be such that $z \in \tilde{\Gamma}_\varepsilon^\phi(x)$. We claim that $z \in \Gamma_\delta^\phi(x)$. In fact, for each $k \in \mathbb{N}$, there exists $c_k \in \mathcal{C}$ such that

$$d(\phi_t(x), \phi_{c_k(t)}(z)) < \frac{3\varepsilon}{2}, \quad \forall -k \le t \le k.$$

Then, for any $-k \le t \le k$,

$$\begin{aligned} &d(\phi_{c_{k+1}(t)-c_k(t)}\phi_{c_k(t)}(z), \phi_{c_k(t)}(z)) \\ &\qquad \le d(\phi_{c_{k+1}(t)}(z), \phi_t(x)) + d(\phi_t(x), \phi_{c_k(t)}(z)) \\ &\qquad < \frac{3\varepsilon}{2} < \gamma. \end{aligned}$$

We observe that $|c_{k+1}(t) - c_k(t)| < \lambda$ for all $-k \le t \le k$. Indeed, suppose not. There is $s \in [-k, k]$ such that $|c_{k+1}(s) - c_k(s)| = \lambda$. Without loss of generality, assume that $c_{k+1}(s) - c_k(s) = \lambda$. Then, by (7.12),

$$\gamma < d(\phi_\lambda \phi_{c_k(s)}(z), \phi_{c_k(s)}(z)) < \gamma.$$

The contradiction shows that $|c_{k+1}(t) - c_k(t)| < \lambda$ for all $-k \le t \le k$.

We now construct a continuous map $c : \mathbb{R} \to \mathbb{R}$ inductively as follows. Set $c(t) = c_1(t)$ for all $t \in [-1, 1]$. Since $|c(1) - c_1(1)| < \lambda$, there exists a continuous function c on $[1, 2]$ such that $c(1) = c_1(1)$, $c_2 = c_2(2)$ and $|c(t) - c_2(t)| < \lambda$ for all $t \in [1, 2]$. Similarly, since $|c(-1) - c_1(-1)| < \lambda$, we can extend c continuously on $[-2, -1]$ such that $c(-1) = c_1(-1)$, $c(-2) = c_2(-2)$ and $|c(t) - c_2(t)| < \lambda$ for all $t \in [-2, -1]$. By this process, we can glue these continuous functions on each interval $[k, k+1]$ for $k \in \mathbb{Z}$ to define a continuous map $c : \mathbb{R} \to \mathbb{R}$ with $c(0) = 0$.

Now fix $t \in \mathbb{R}$ and assume $t > 0$. We consider the following two cases.

Case 1 $t \in [0, 1]$. We have

$$d(\phi_t(x), \phi_{c(t)}(z)) = d(\phi_t(x), \phi_{c_1(t)}(z)) < \frac{3\varepsilon}{2} = \frac{\delta}{2m} < \delta.$$

Case 2 $t \in [k, k+1]$ for some $k \geq 1$. Since $|c(t) - c_{k+1}(t)| < \lambda$, it follows from (7.11) that

$$d(\phi_{c(t)}(z), \phi_{c_{k+1}(t)}(z)) < \frac{\delta}{2}.$$

Therefore,

$$\begin{aligned} d(\phi_t(x), \phi_{c(t)}(z)) &\leq d(\phi_t(x), \phi_{c_{k+1}(t)}(z)) + d(\phi_{c_{k+1}(t)}(z), \phi_{c(t)}(z)) \\ &\leq \frac{3\varepsilon}{2} + \frac{\delta}{2} < \delta. \end{aligned}$$

Hence $z \in \Gamma^{\phi}_{\delta}(x)$, and consequently $\tilde{\Gamma}^{\phi}_{\varepsilon}(x) \subset \Gamma^{\phi}_{\delta}(x)$.

Finally, by definition, we have $\Gamma^{\phi}_{\varepsilon}(x) \subset \tilde{\Gamma}^{\phi}_{\varepsilon}(x)$, which completes the proof. □

Lemma 7.9 *Let ϕ be a continuous flow without singularities on a compact metric space X. Then, for every sufficiently small $\lambda > 0$, there exists $\varepsilon > 0$ such that the following holds: for all $x, y \in X$, all interval $[0, T] \subset \mathbb{R}$, and all $\alpha \in \mathcal{C}$, if*

$$d(\phi_t(x), \phi_{\alpha(t)}(y)) \leq \varepsilon, \quad \forall t \in [0, T],$$

then

$$|\alpha(t) - t| \leq \lambda \text{ for } t \leq 2, \quad \text{and} \quad |\alpha(t) - t| \leq t\lambda \text{ for } t > 2,$$

for all $t \in [0, T]$.

Proof Let T_0 be as in Lemma 6.4, and $0 < \lambda < T_0$. For $\lambda > 0$, take $\gamma > 0$ such that

$$d(\phi_\lambda(x), x) > \gamma, \quad \forall x \in X. \tag{7.13}$$

By the continuity of ϕ, take $0 < \varepsilon < \gamma$ such that if $d(x, y) \leq \varepsilon$, then

$$d(\phi_t(x), \phi_t(y)) < \gamma \quad \text{for all } 0 \leq t \leq 2.$$

Suppose that

$$d(\phi_t(x), \phi_{\alpha(t)}(y)) \le \varepsilon, \quad \forall t \in [0, 2].$$

for some $x, y \in X$ and $\alpha \in \mathcal{C}$.

We first claim that $|\alpha(t) - t| < \lambda$ for all $t \in [0, 2]$. Indeed, suppose not. Then there exists $t_0 \in [0, 2]$ such that $|\alpha(t_0) - t_0| = \lambda$. Without loss of generality, assume $\alpha(t_0) > t_0$. Since $d(x, y) \le \varepsilon$, we have $d(\phi_{t_0}(x), \phi_{t_0}(y)) < \gamma$. By (7.13), we get

$$\gamma < d(\phi_{t_0}(x), \phi_\lambda(\phi_{t_0}(y))) = d(\phi_{t_0}(x), \phi_{\alpha(t_0)}(y)) \le \varepsilon.$$

This contradiction proves the claim.

Next we consider the case $t \in [2, 4]$ and suppose that

$$d(\phi_t(x), \phi_{\alpha(t)}(y)) \le \varepsilon.$$

Let $u = t - 2$. Then

$$d(\phi_u(\phi_2(x)), \phi_{\alpha(u+2)-\alpha(2)}(\phi_{\alpha(2)}(y))) = d(\phi_{u+2}(x), \phi_{\alpha(u+2)}(y)) \le \varepsilon.$$

Define $\beta : [0, 2] \to \mathbb{R}$ by $\beta(u) = \alpha(u + 2) - \alpha(2)$. Clearly, $\beta(0) = 0$. By the claim above, we get $|\beta(u) - u| < \lambda$ for all $u \in [0, 2]$. Hence,

$$\begin{aligned} |\alpha(u + 2) - (u + 2)| &= |\beta(u) - u + \alpha(2) - 2| \\ &\le |\beta(u) - u| + |\alpha(2) - 2| \\ &< 2\lambda. \end{aligned}$$

It follows that $|\alpha(t) - t| \le 2\lambda$.

By repeating the same argument, we can show that for $t \in [2n - 2, 2n]$ with $n \ge 1$,

$$|\alpha(t) - t| < n\lambda.$$

Consequently, for each $t > 2$ in $[0, T]$, there is $n \ge 2$ such that $t \in [2n - 2, 2n]$. Then, we obtain

$$|\alpha(t) - t| \le n\lambda = \frac{n}{t} t\lambda \le t\lambda.$$

□

Lemma 7.10 *Let ϕ be a continuous flow on a compact metric space X without fixed points. There exists $\varepsilon > 0$ such that for $x \in X$, $r > 0$ and a sequences $\{c_n\}_{n\in\mathbb{N}}$ in $\mathcal{C}$ and $\{y_n\}_{n\in\mathbb{N}}$ in X with $y_n \to y$ for some $y \in X$ satisfying*

$$d(\phi_t(x), \phi_{c_n(t)}(y_n)) \le \varepsilon, \quad \forall n \in \mathbb{N} \textit{ and } t \in [-r, r],$$

then for every $\delta > 0$ there is $N \in \mathbb{N}$ satisfying

$$d(\phi_{c_n(t)}(y_n), \phi_{c_n(t)}(y)) \le \delta, \quad \forall t \in [-r, r] \text{ and } n \ge N.$$

Proof Let $T_0 > 0$ be as in Lemma 6.4, and $0 < \lambda < T_0$. Take $\varepsilon > 0$ such that

$$d(\phi_\lambda(x), x) > \varepsilon, \quad \forall x \in X.$$

Let $x \in X, r > 0$ and a sequences $\{c_n\}_{n\in\mathbb{N}}$ in $\mathcal{C}$ and $\{y_n\}_{n\in\mathbb{N}}$ in X with $y_n \to y$ for some $y \in X$ satisfying $d(\phi_t(x), \phi_{c_n(t)}(y_n)) \le \varepsilon$ for all $n \in \mathbb{N}$ and $t \in [-r, r]$. Suppose, on the contrary, that there is a constant $\delta > 0$, and for each $k \in \mathbb{N}$, there are $n_k > k$ and $t_k \in [-r, r]$ such that

$$d(\phi_{c_{n_k}(t_k)}(y_{n_k}), \phi_{c_{n_k}(t_k)}(y)) > \delta. \tag{7.14}$$

By Lemma 7.9, we observe that $|c_{n_k}(t_k) - t_k| < \lambda \max\{|t_k|, 1\}$. Since $t_k \in [-r, r]$ for all $k \in \mathbb{N}$, it follows that $\{c_{n_k}(t_k)\}_{k\in\mathbb{N}}$ is bounded. We may assume that $c_{n_k}(t_k) \to s$ for some $s \in \mathbb{R}$. By letting $k \to \infty$ in (7.14), we get a contradiction. This completes the proof.

□

In the following lemma, we show that the set $\tilde{\Gamma}^\phi_\delta(x)$ is compact if δ is sufficiently small. For simplicity, for $x \in X, \gamma, r > 0$, we define

$$B^\phi_r[x, \gamma] = \bigcup_{c\in\mathcal{C}} \bigcap_{|t|\le r} \phi_{-c(t)}(B[\phi_t(x), \gamma]).$$

Lemma 7.11 *Let ϕ be a flow on a compact metric space X without fixed points. There exists $\delta_0 > 0$ such that for every $\delta < \delta_0$ and $x \in X$, $\tilde{\Gamma}^\phi_\delta(x)$ is compact.*

Proof Let $T_0 > 0$ be as in Lemma 6.4, and fix $0 < \lambda < T_0$. Choose $\delta_0 > 0$ such that

$$d(\phi_\lambda(x), x) > \delta_0, \quad \forall x \in X.$$

Let $0 < \delta < \delta_0$. Since $B^\phi_r[x, \gamma_1] \subset B^\phi_r[x, \gamma_2]$ for $\gamma_1 > \gamma_2$, it is sufficient to show that for any $x \in X, \delta < \gamma < \delta_0$ and $r > 0$, the set $B^\phi_r[x, \gamma]$ is closed.

Fix $\beta > 0$ such that $\delta < \beta < \gamma$, and let $\{y_n\}_{n\in\mathbb{N}}$ be a sequence in $B^\phi_r[x, \gamma]$ with $y_n \to y$ for some $y \in X$. Then, for each $n \in \mathbb{N}$, there exists $c_n \in \mathcal{C}$ such that

$$d(\phi_t(x), \phi_{c_n(t)}(y_n)) \le \beta, \quad \forall t \in [-r, r]. \tag{7.15}$$

Moreover, by Lemma 7.10, there exists $N \in \mathbb{N}$ such that

$$d(\phi_{c_n(t)}(y_n), \phi_{c_n(t)}(y)) \le \gamma - \beta, \quad \forall t \in [-r, r] \text{ and } n \ge N. \tag{7.16}$$

By (7.15) and (7.16), we obtain

$$\begin{aligned} d(\phi_t(x), \phi_{c_n(t)}(y)) &\leq d(\phi_t(x), \phi_{c_n(t)}(y_n)) + d(\phi_{c_n(t)}(y_n), \phi_{c_n(t)}(y)) \\ &\leq \gamma, \end{aligned}$$

and so $y \in B_r^{\phi}[x, \gamma]$. Therefore, $B_r^{\phi}[x, \gamma]$ is closed, completing the proof. □

Lemma 7.12 *Let ϕ be a flow on a compact metric space X, and let A be a closed subset of X which intersects uncountably many orbits of ϕ. Then there exists a ϕ-orbit vanishing measure on X supported on A.*

Proof Let Y denote the set of all points $a \in A$ such that any neighborhood of a in X intersects uncountably many orbits of ϕ. We first show that Y is nonempty. Indeed, if Y is empty, then A can be covered by finitely many open sets, each intersecting only countably many orbits of ϕ, which contradicts the definition of A. Hence, Y is nonempty.

Next, we show that Y is compact. Let $\{y_n\}_{n\in\mathbb{N}}$ be a sequence in Y such that $y_n \to y$ for some $y \in X$. For any neighborhood U in X of y, there exists $y_n \in U$. Since U intersects uncountably many orbits of ϕ, it follows that $y \in Y$. Thus, Y is compact.

We now prove that there exists a ϕ-orbit vanishing measure supported on Y. Suppose, on the contrary, that no ϕ-orbit vanishing measure exists on Y. For each $n \in \mathbb{N}$, define

$$C(\tfrac{1}{n}) = \left\{\mu \in \mathcal{M}(Y) \;\middle|\; \exists x \in Y \text{ such that } \mu\big(\phi_{[-n,n]}(x) \cap Y\big) \geq \tfrac{1}{n}\right\}.$$

Then, for any measure $\mu \in \mathcal{M}(Y)$, there exists $x \in Y$ such that $\mu(\phi_{\mathbb{R}}(x) \cap Y) > 0$. Hence, we can choose $n \in \mathbb{N}$ such that

$$\mu(\phi_{[-n,n]}(x) \cap Y) > \frac{1}{n}.$$

Otherwise,

$$\mu(\phi_{\mathbb{R}}(x) \cap Y) = \lim_{n\to\infty} \mu(\phi_{[-n,n]}(x) \cap Y) = 0,$$

which is a contradiction. Thus $\mu \in C(1/n)$ for some $n \in \mathbb{N}$, and therefore $\mathcal{M}(Y) = \bigcup_{n=1}^{\infty} C(1/n)$.

We claim that $C(\frac{1}{n})$ is closed for each $n \in \mathbb{N}$. Indeed, let $\{\mu_m^n\}_{m=1}^{\infty}$ be a sequence in $C(\frac{1}{n})$ with $\mu_m^n \to \mu^n$ as $m \to \infty$. Since each μ_m^n is an element of $C(\frac{1}{n})$, there is $x_m^n \in Y$ such that

$$\mu_m^n(\phi_{[-n,n]}(x_m^n) \cap Y) \geq \frac{1}{n}.$$

Because Y is compact, we may assume $x_m^n \to x^n$ as $m \to \infty$ for some $x^n \in Y$. For any closed neighborhood C of $\phi_{[-n,n]}(x^n)$ in X, there is $M > 0$ such that $\phi_{[-n,n]}(x_m^n) \subset C$ for all $m \geq M$. Since μ_m^n converges to μ^n, we have

$$\mu^n(C \cap Y) \geq \limsup_{m\to\infty} \mu_m^n(C \cap Y) \geq \limsup_{m\to\infty} \mu_m^n(\phi_{[-n,n]}(x_m^n) \cap Y) \geq \frac{1}{n}.$$

Then $\mu^n(\phi_{[-n,n]}(x^n) \cap Y) \geq \frac{1}{n}$, and so $\mu^n \in C(\frac{1}{n})$.

Since Y is compact, Theorem 1.4 implies that $\mathcal{M}(Y)$ is also compact. By Baire's category theorem, there exists $n_0 \in \mathbb{N}$ such that $C(\frac{1}{n_0})$ has nonempty interior; that is, $intC(\frac{1}{n_0}) \neq \emptyset$. By Theorem 1.5, we note that the collection of measure $\mu \in \mathcal{M}(Y)$ with finite support is dense in $\mathcal{M}(Y)$. Hence there exists a Borel measure μ on Y with finite support such that $\mu_0 \in intC(\frac{1}{n_0})$. We assume that $supp(\mu_0) = \{p_1, \ldots p_h\}$ and μ_0 have masses $q_1, \ldots, q_h$ at the points $p_1, \ldots, p_h$, respectively. For each $j = 1, \ldots, h$, let $\{x_k^j\}_{k=1}^{\infty}$ be a sequence converging to p_j such that $x_k^j \in B(p_j, 1/k) \cap Y$ and x_k^i and x_l^j belong to different orbits if $k \neq l$ and $i \neq j$. Let $N > n_0$ be such that

$$a := \frac{1}{N}\left(1 - \frac{1}{2^N}\right) < \frac{1}{n_0}.$$

For each $k \in \mathbb{N}$, we define a measure μ_k by

$$\mu_k(B) = \sum_{j=1}^{h}\left(\sum_{i=1}^{N} q_j a\, \delta_{x_{k+i}^j}(B) + \sum_{i=N+1}^{\infty} \frac{q_j}{2^i}\delta_{x_{k+i}^j}(B)\right),$$

where B is a Borel subset of Y and δ_x is the Dirac measure centered at x. Clearly, μ_k is a Borel probability measure on Y, and $\mu_k(\phi_{\mathbb{R}}(x) \cap Y) < \frac{1}{n_0}$ for all $x \in Y$.

We claim that $\mu_k \to \mu_0$ as $k \to \infty$. Indeed, let U be an open subset of Y. If $p_j \notin U$ for all $j = 1, \ldots, h$, then

$$\liminf_{k\to\infty} \mu_k(U) \geq \mu_0(U) = 0.$$

Otherwise, suppose $\{p_1, \ldots, p_t\} \subset U$ for some $1 \leq t \leq h$. Then there exists $N > 0$ such that $x_n^j \in U$ for all $n \geq N$ and $j = 1, \ldots, t$, and hence

$$\sum_{j=1}^{t} q_j = \liminf_{k\to\infty} \mu_k(U) \geq \mu_0(U) = \sum_{j=1}^{t} q_j.$$

By Theorem 1.3, we conclude that $\mu_k \to \mu_0$ as $k \to \infty$.

Since $\mathrm{int}C(\frac{1}{n_0})$ is a neighborhood of μ_0, there exists $\mu_k \in \mathrm{int}C(\frac{1}{n_0})$. However, by construction,

$$\mu_k(\phi_{[-n_0,n_0]}(x)) < \frac{1}{n_0}, \quad \forall x \in Y,$$

which contradicts the definition of $C(\frac{1}{n_0})$. Therefore, there exists a ϕ-orbit vanishing measure on Y.

□

Corollary 7.2 *Let ϕ be a flow on a compact metric space X. Suppose that any nonempty open subset of X intersects uncountably many orbits of ϕ. Then $\mathcal{M}_\phi^*(X)$ is dense in $\mathcal{M}(X)$.*

Proof As in the proof of Lemma 7.12 with $A = X$, for each $n \in \mathbb{N}$, define

$$C(\tfrac{1}{n}) = \{\mu \in \mathcal{M}(X) \mid \exists x \in X \text{ such that } \mu(\phi_{[-n,n]}(x)) \geq \tfrac{1}{n}\}.$$

It follows that $C(\frac{1}{n})$ is closed and nowhere dense in $\mathcal{M}(X)$ for every $n \in \mathbb{N}$. Consequently, $\mathcal{M}(X) \setminus C(\frac{1}{n})$ is open and dense in $\mathcal{M}(X)$. Since $\mathcal{M}(X)$ is a Baire space, we obtain

$$\mathcal{M}_\phi^*(X) = \mathcal{M}(X) \setminus \bigcup_{n=1}^{\infty} C(\tfrac{1}{n}) = \bigcap_{n=1}^{\infty} (\mathcal{M}(X) \setminus C(\tfrac{1}{n})).$$

is dense in $\mathcal{M}(X)$. This completes the proof.

□

Proof of Theorem 7.4 Suppose ϕ is measure expansive. By Proposition 7.1, we can choose a uniform expansive constant $\delta > 0$ such that $\mu(\Gamma_{2\delta}^\phi(x)) = 0$ for every $\mu \in \mathcal{M}_\phi^*(X)$ and $x \in X$.

Case 1 Suppose ϕ has no fixed points. For the constant $\delta > 0$ given above, Lemmas 7.8 and 7.11 ensures the existence of a sufficiently small constant $\varepsilon > 0$ such that $\tilde{\Gamma}_\varepsilon^\phi(x)$ is compact and $\Gamma_\varepsilon^\phi(x) \subset \tilde{\Gamma}_\varepsilon^\phi(x) \subset \Gamma_\delta^\phi(x)$ for all $x \in X$. Suppose, on the contrary, that ϕ is not countably expansive. Then, for each $n \in \mathbb{N}$, there exists $x_n \in X$ such that $\Gamma_{1/n}^\phi(x_n)$ intersect with uncountably many orbits of ϕ. Consequently, the compact set $\tilde{\Gamma}_{1/n}^\phi(x_n)$ also meets uncountably many orbits of ϕ. By Lemma 7.12, we can find a ϕ-orbit vanishing measure μ_n supported on $\tilde{\Gamma}_{1/n}^\phi(x_n)$ for each $n \in \mathbb{N}$. Define a Borel measure μ on X by

$$\mu(A) = \sum_{n=1}^{\infty} \frac{\mu_n(A)}{2^n}$$

for all Borel subset A of X. It is clear that μ is ϕ-orbit vanishing. Take $n \in \mathbb{N}$ with $1/n < \varepsilon$. Then

$$\mu(\Gamma_\delta^\phi(x_n)) \geq \mu(\tilde{\Gamma}_\varepsilon^\phi(x_n)) \geq \frac{\mu_n(\tilde{\Gamma}_{1/n}^\phi(x_n))}{2^n} > 0,$$

contradicting the measure expansiveness of ϕ. Hence, ϕ is countably expansive.

Case 2 Suppose ϕ has at least one fixed point. Let $U_\delta(Fix(\phi))$ denote the δ-neighborhood of the collection of fixed points $Fix(\phi)$.

Claim 1: We claim that for any $p \in Fix(\phi)$, the set $\Gamma_{2\delta}^{\phi}(p)$ intersects at most countably many orbits of ϕ. Indeed, suppose, on the contrary, that there exists $p \in Fix(\phi)$ such that $\Gamma_{2\delta}^{\phi}(p)$ intersects with uncountably many orbits of ϕ. Since $\Gamma_{2\delta}^{\phi}(p) = \{y \in X : d(p, y) \leq 2\delta\}$, by Lemma 7.12, there is a Borel measure $\mu \in \mathcal{M}_{\phi}^{*}(X)$ such that $\mu(\Gamma_{2\delta}^{\phi}(p)) = 1$, which contradicts the measure expansiveness of ϕ. This contradiction proves Claim 1.

Claim 2: There exists $\varepsilon > 0$ such that, for every $x \in X$, the set $\Gamma_{\varepsilon}^{\phi}(x)$ intersects at most countably many orbits of ϕ. Let $Y = \{x \in X : \phi_{\mathbb{R}}(x) \subset X \setminus U_{\delta}(Fix(\phi))\}$. For any $x \in Y^c$, there exist $t \in \mathbb{R}$ and $p \in Fix(\phi)$ such that $\phi_t(x) \in B(p, \delta)$. By Claim 1, Y^c consists of countably many orbits of ϕ. Without loss of generality, we may assume that Y is nonempty. Clearly, Y is closed and ϕ-invariant. Let $\psi = \phi|_Y$ denote the restriction of the flow ϕ on Y. It is clear that ψ has no fixed points. Moreover, for each $x \in Y$, $\Gamma_{\delta}^{\psi}(x) = \Gamma_{\delta}^{\phi}(x) \cap Y$.

We now define a map $i : \mathcal{M}_{\psi}^{*}(Y) \to \mathcal{M}_{\phi}^{*}(X)$ by

$$i(\mu)(A) = \mu(A \cap Y), \quad \forall \mu \in \mathcal{M}_{\psi}^{*}(Y),$$

for every Borel subset A of X and every $\mu \in \mathcal{M}_{\psi}^{*}(Y)$. Since ϕ is measure expansive, we have

$$\mu(\Gamma_{\delta}^{\psi}(x)) = \mu(\Gamma_{\delta}^{\phi}(x) \cap Y) = i(\mu)(\Gamma_{\delta}^{\phi}(x)) = 0$$

for all $\mu \in \mathcal{M}_{\psi}^{*}(Y)$ and $x \in Y$. Hence ψ is measure expansive. By Case 1, there exists $\varepsilon > 0$ such that, for any $x \in Y$, $\Gamma_{\varepsilon}^{\psi}(x)$ intersects at most countably many orbits of ψ. For any $x \in Y$, we observe that

$$\Gamma_{\varepsilon}^{\phi}(x) = \Gamma_{\varepsilon}^{\psi}(x) \cup (\Gamma_{\varepsilon}^{\phi}(x) \cap Y^c).$$

Since Y^c consists of countably many orbits of ϕ, it follows that $\Gamma_{\varepsilon}^{\phi}(x)$ intersects at most countably many orbits of ϕ for all $x \in Y$. Moreover, for any $x \in Y^c$, we can suppose that $x \in B(p, \delta)$ for some $p \in Fix(\phi)$, and so $\Gamma_{\varepsilon}^{\phi}(x) \subset \Gamma_{2\delta}^{\phi}(p)$. By Claim 1, we obtain that $\Gamma_{\varepsilon}^{\phi}(x)$ intersects at most countably many orbits of ϕ. Therefore, ϕ is countably expansive.

On the other hand, it is clear that every countably expansive flow is measure expansive. This complete the proof.

□

Chapter 8
Shadowing and Topological Stability for Flows

In this chapter, we study the shadowing and topologically stable measures for a flow on a compact metric space X. In Sect. 8.1, we introduce a notion of a shadowing measure for a flow on X, which is applied to characterize the shadowing flows. In Sect. 8.2, we introduce a concept of a topologically stable measure for a flow on X. Moreover, we present a measurable version of Walters' stability theorem for flows which says that any expansive measure with shadowing is topologically stable.

8.1 Shadowing Measures for Flows

In this section, we introduce the notions of shadowing measures and shadowable points for flows on compact metric spaces, and use them to characterize the notion of shadowing for flows. Throughout this section, we assume that ϕ is a flow on a compact metric space X, and that $\mu \in \mathcal{M}(X)$. A $(\delta, 1)$-pseudo orbit $\xi = \{(x_i, t_i)\}_{i\in\mathbb{Z}}$ of ϕ is said to be through a subset B of X if $x_0 \in B$. In particular, we say that ξ is through $x \in X$ if $B = \{x\}$. Motivated by the notion of shadowing measures for homeomorphisms, we introduce a notion of shadowing measures for flows as follows.

Definition 8.1 A Borel measure $\mu \in \mathcal{M}(X)$ is said to be *shadowing* for ϕ on an invariant subset Λ of X if for any $\varepsilon > 0$ there are $\delta > 0$ and a Borel set $B \subset \Lambda$ with $\mu(X \setminus B) = 0$ such that every $(\delta, 1)$-pseudo orbit $\xi = \{(x_i, t_i)\}_{i\in\mathbb{Z}}$ in Λ through B is ε-shadowed by a point in X.

Definition 8.2 A Borel measure $\mu \in \mathcal{M}(X)$ is said to be *invariant* for ϕ if $\mu(A) = \mu(\phi_t(A))$ for all $t \in \mathbb{R}$. We denote by $\mathcal{M}(X, \phi)$ the collection of all invariant measures for ϕ.

Definition 8.3 We say that ϕ is *measure shadowing* (resp. *invariant measure shadowing*) on an invariant subset Λ if μ is shadowing for $\phi|_\Lambda$ for all $\mu \in \mathcal{M}(\Lambda)$ (resp. $\mu \in \mathcal{M}(\Lambda, \phi|_\Lambda)$). If μ is shadowing for ϕ on X then we say that μ is shadowing for ϕ.

K. Lee et al., *Topological Dynamics from a Measure-Theoretic Viewpoint*, Springer Asia Pacific Mathematics Series 14, https://doi.org/10.1007/978-981-95-6390-6_8

In the following proposition, we give a characterization of shadowing measures for flows using their support. For this we need an auxiliary concept as follows.

Definition 8.4 We say that a flow ϕ is *shadowing* through a subset B of X if for any $\varepsilon > 0$, there is $\delta > 0$ such that every $(\delta, 1)$-pseudo orbit through B is ε-shadowed by a point in X.

Proposition 8.1 *Let ϕ be a flow on a compact metric space X. Then a Borel measure $\mu \in \mathcal{M}(X)$ is shadowing if and only if ϕ is shadowing through* $\operatorname{supp}(\mu)$.

Proof If ϕ is shadowing through $\operatorname{supp}(\mu)$, μ is shadowing for ϕ by taking $B = \operatorname{supp}(\mu)$ in the corresponding definition. Conversely, suppose that μ is shadowing for ϕ. Let $\varepsilon > 0$. We take a constant $0 < \delta < \varepsilon/2$ and a Borel subset B of X such that $\mu(B) = 1$ and any 2δ-pseudo orbit of ϕ through B can be $\frac{\varepsilon}{2}$-shadowed.

Let $\{(x_n, t_n)\}_{n\in\mathbb{Z}}$ be a $(\delta, 1)$-pseudo orbit through $\operatorname{supp}(\mu)$. Since $\mu(B) = 1$, we have $\operatorname{supp}(\mu) \subset \overline{B}$, and so there is $y \in B$ such that

$$d(y, x_0) < \delta \quad \text{and} \quad d(\phi_{t_0}(y), \phi_{t_0}(x_0)) < \delta.$$

Then, the sequence $\xi = \{(y_n, t_n)\}_{n\in\mathbb{Z}}$ given by

$$y_n = \begin{cases} y & \text{if } n = 0, \\ x_n & \text{otherwise} \end{cases}$$

is a $(2\delta, 1)$-pseudo orbit of ϕ through B. Hence, there are $z \in X$ and $h \in \mathcal{H}$ such that

$$d(\phi_{h(t)}(z), \phi_{t-T_n}(y_n)) < \frac{\varepsilon}{2}, \quad \forall t \in [T_n, T_{n+1}) \text{ and } \quad \forall n \in \mathbb{Z},$$

where T_n is the nth sum associated to ξ. Since

$$d(z, x_0) \leq d(z, y) + d(y, x_0) < \varepsilon,$$

we derive that ξ is ε-shadowed by $z \in X$. Hence, ϕ is shadowing through $\operatorname{supp}(\mu)$. □

Definition 8.5 A point $p \in X$ is said to be *shadowable* for ϕ if for every $\varepsilon > 0$ there exists $\delta > 0$ such that every $(\delta, 1)$-pseudo-orbit through $\{p\}$ is ε-shadowed by a point in X. We denote by $Sh(\phi)$ the set of all shadowable points of ϕ in X.

The following proposition establishes that the set $Sh(\phi)$ is invariant under the flow ϕ. The proof can be found in [4].

Proposition 8.2 *Let ϕ be a flow on a compact metric space X. Then $Sh(\phi)$ is invariant.*

In the following, we present the relationship between the shadowable points of a homeomorphism and those of its suspension flow. The result is based on Theorem 2.7 in [4]. To establish it, we first recall two technical lemmas.

Lemma 8.1 *Let f be a homeomorphism on a compact metric space X, and ϕ^f be the suspension flow of f under the constant height function $\tau = 1$. If $d^f((x,s),(y,t)) < \varepsilon < \frac{1}{4}$ for $(x,s),(y,t) \in X^f$ then $|s-t| < \varepsilon$ or $|1+s-t| < \varepsilon$ or $|1+t-s| < \varepsilon$.*

Proof See Lemma 2.4 in [37].

Lemma 8.2 *Let f be a homeomorphism on a compact metric space X, and ϕ^f be the suspension flow of f under the constant height function $\tau = 1$. Then for all $\varepsilon > 0$, there is $\delta > 0$ such that if $d^f((x,s),(y,t)) < \delta$ for $(x,s),(y,t) \in X^f$, then $d(x,y) < \varepsilon$ or $d(f(x),y) < \varepsilon$ or $d(f^{-1}(x),y) < \varepsilon$.*

Proof See Lemma 2.5 in [37].

Theorem 8.1 *Let f be a homeomorphism on a compact metric space X, and ϕ^f be the suspension flow of f under the height function $\tau = 1$. Then*

$$Sh(\phi^f) = (Sh(f) \times [0,1])/\sim.$$

Proof Let $\overline{z} = (z,t) \in Sh(\phi^f)$. We claim that $z \in Sh(f)$. Since $Sh(\phi^f)$ is invariant, we may assume that $\overline{z} = (z, \frac{1}{2}) \in Sh(\phi^f)$. Fix $\varepsilon > 0$, and choose $0 < \varepsilon' < \min\{\varepsilon, \frac{1}{4}\}$ such that

$$\text{if } d(x,y) < \varepsilon \text{ then } d(f^i(x), f^i(y)) < \varepsilon \text{ for } i \in \{-1,0,1\}. \tag{8.1}$$

Since $\overline{z} \in Sh(\phi^f)$, there exists $0 < \delta < \varepsilon'$ such that every $(\delta,1)$-pseudo orbit of ϕ^f through $\overline{z}$ is ε'-shadowed by some point in X^f. By continuity of f, choose $0 < \delta' < \delta$ such that $d(x,y) < \delta'$ implies $d(f(x),f(y)) < \delta$. Let $\xi = \{x_n\}_{n\in\mathbb{Z}}$ be a δ'-pseudo orbit of f through z, and consider the sequence $\{((x_n, \frac{1}{2}),1)\}_{n\in\mathbb{Z}}$ in X^f. Observe that

$$\begin{aligned} d^f\left(\phi^f_{t_n}\left(x_n, \tfrac{1}{2}\right), \left(x_{n+1}, \tfrac{1}{2}\right)\right) &= d^f\left(\left(f(x_n), \tfrac{1}{2}\right), \left(x_{n+1}, \tfrac{1}{2}\right)\right) \\ &= \tfrac{1}{2} d(f(x_n), x_{n+1}) + \tfrac{1}{2} d(f^2(x_n), f(x_{n+1})) \le \delta. \end{aligned}$$

Hence, $\left\{\left(\left(x_n, \frac{1}{2}\right), 1\right)\right\}_{n\in\mathbb{Z}}$ is a $(\delta,1)$-pseudo orbit of ϕ^f through $\overline{z}$. By the choice of δ, there exist $(x,s) \in X^f$ and $h \in \mathcal{H}$ such that

$$d^f(\phi^f_{h(t)}(x,s), \phi^f_{t-n}(x_n, \frac{1}{2})) < \varepsilon' \text{ for } n \le t < n+1 \text{ and } n \in \mathbb{Z}.$$

For $t = 0$, we obtain

$$d^f\left((x,s), \left(z, \tfrac{1}{2}\right)\right) < \frac{1}{4},$$

which implies $\left|s - \frac{1}{2}\right| < \frac{1}{4}$. Moreover, since

$$d^f\left(\phi^f_{h(t)}(x,s), \phi^f_t\left(z, \tfrac{1}{2}\right)\right) < \varepsilon' < \tfrac{1}{4}, \quad \forall 0 \le t < 1,$$

we have $1 \leq s + h(1) < 2$, and so we may write $\phi^f_{h(1)}(x, s) = (f(x), s^{(1)})$ for some $0 \leq s^{(1)} < 1$. Similarly, since

$$d^f\left(\phi^f_{h(t)}(x, s), \phi^f_{t-1}(x_1, \tfrac{1}{2})\right) < \frac{1}{4}, \quad \forall 1 \leq t < 2,$$

we can write $\phi^f_{h(2)}(x, s) = (f^2(x), s^{(2)})$ for some $0 \leq s^{(2)} < 1$. Continuing this process, for each $n \in \mathbb{Z}$, we have $\phi^f_{h(n)}(x, s) = (f^n(x), s^{(n)})$ for some $0 \leq s^{(n)} < 1$.

If $f^n(x) = x_n$, then it is obvious that $d(f^n(x), x_n) < \varepsilon$. Otherwise, we have

$$\tfrac{1}{2}d(f^n(x), x_n) + \tfrac{1}{2}d(f^{n+1}(x), f(x_n)) \leq d^f((f^n(x), s^{(n)}), (x_n, \tfrac{1}{2})) < \varepsilon'.$$

Thus, $d(f^n(x), x_n) < \varepsilon'$ or $d(f^{n+1}(x), f(x_n)) < \varepsilon'$ for all $n \in \mathbb{Z}$. By (8.1), we then have $d(f^n(x), x_n) < \varepsilon$ for every $n \in \mathbb{Z}$. Therefore, ξ is ε-shadowed by x, and hence $z \in Sh(f)$.

Conversely, let $z \in Sh(f)$ and $r \in [0, 1]$. We claim that $(z, r) \in Sh(\phi^f)$. Let $\varepsilon > 0$ and choose $0 < \varepsilon' < \varepsilon$ such that $d(x, y) < \varepsilon'$ implies $d(f^i(x), f^i(y)) < \frac{\varepsilon}{2}$ for $i \in \{0, 1, 2\}$. Since $z \in Sh(f)$, there exists $0 < \delta < \frac{\varepsilon'}{2}$ such that every δ-pseudo orbit of f through z is ε'-shadowed by some point in X. Choose $0 < \delta' < \min\{\frac{1}{4}, \delta\}$ corresponding to δ as in Lemma 8.2. Let $\xi = \{(x_k, s_k), t_k\}_{k \in \mathbb{Z}}$ be a $(\delta', 2, 4)$-pseudo orbit of ϕ^f on X^f such that $(x_0, s_0) = (z, r)$. Let $w_k = [s_k + t_k]$ denote the integer part of $s_k + t_k$. We have

$$d^f((f^{w_k}(x_k), s_k + t_k - w_k), (x_{k+1}, s_{k+1})) < \delta' \quad \text{for all } k \in \mathbb{Z}.$$

Since $\delta' < \frac{1}{4}$, Lemma 8.1 implies that one of the following inequalities holds:

$$|s_k + t_k - w_k - s_{k+1}| < \delta' \quad \text{or} \quad |1 + s_k + t_k - w_k - s_{k+1}| < \delta'$$
$$\text{or} \quad |1 + s_{k+1} + w_k - t_k - s_k| < \delta'.$$

Now, let n_k be a positive integer defined as follows:

$$n_k = \begin{cases} w_k, & \text{if } |s_k + t_k - w_k - s_{k+1}| < \delta', \\ w_k - 1, & \text{if } |1 + s_k + t_k - w_k - s_{k+1}| < \delta', \\ w_k + 1, & \text{if } |1 + s_{k+1} + w_k - t_k - s_k| < \delta'. \end{cases}$$

Then, by Lemma 8.2, we obtain that $d(f^{n_k}(x_k), x_{k+1}) < \delta$ for all $k \in \mathbb{Z}$. Consider a sequence $\{y_i\}_{i \in \mathbb{Z}}$ in X given by

$$y_i = f^{i - N_k}(x_k) \quad \text{for } N_k \leq i < N_{k+1},$$

where

$$N_k = \begin{cases} 0, & \text{if } k = 0, \\ \sum_{i=0}^{k-1} n_i, & \text{if } k > 0, \\ \sum_{i=-1}^{k} n_i, & \text{if } k < 0. \end{cases}$$

It follows directly from the construction that $\{y_i\}_{i\in\mathbb{Z}}$ is a δ-pseudo orbit of f through z. Since $z \in Sh(f)$, there exists $x \in X$ such that $d(f^i(x), y_i) < \varepsilon'$ for all $i \in \mathbb{Z}$. In particular, we have

$$d(f^{j+N_k}(x), f^j(x_k)) < \varepsilon' \quad \text{for } 0 \le j < n_k \ (k \in \mathbb{Z}). \tag{8.2}$$

Define a function $g : \mathbb{R} \to \mathbb{R}$ by

$$g(t) = \frac{s_{k+1} + n_k - s_k}{t_k}(t - T_k) + s_k + N_k - s_0, \quad \text{for } T_k \le t < T_{k+1},$$

where T_k is the kth sum associated to ξ. It is clear that $g \in \mathcal{H}$. We now observe that the sequence $((x_k, s_k), t_k)_{k\in\mathbb{Z}}$ is ε-shadowed by (x, r). Indeed, let $t \in \mathbb{R}$, and choose $k \in \mathbb{Z}$ such that $T_k \le t < T_{k+1}$. Then we get

$$\begin{aligned} \left|g(t) - s_k - N_k + s_0 - (t - T_k)\right| &= \left|\frac{s_{k+1} + n_k - s_k - t_k}{t_k}(t - T_k)\right| \\ &= |s_{k+1} + n_k - s_k - t_k| \left|\frac{t - T_k}{t_k}\right|. \end{aligned}$$

Since $|s_k + t_k - n_k - s_{k+1}| < \delta'$ and $0 \le t - T_k < t_k$, it follows that

$$\left|g(t) - s_k - N_k + s_0 - (t - T_k)\right| < \delta'.$$

Let j be a positive integer such that $0 \le s_k + t - T_k - j < 1$. Then $0 \le j \le s_k + t_k \le n_k + 2$. By (8.2) and the choice of ε', we obtain

$$d(f^{j+N_k}(x), f^j(x_k)) < \tfrac{\varepsilon}{2} \quad \text{for } 0 \le j \le n_k + 2.$$

Finally, we have

$$\begin{aligned} & d^f(\phi^f_{g(t)}(x, r), \phi^f_{t-T_k}(x_k, s_k) \\ &= d^f\left((f^{N_k}(x), s_0 + g(t) - N_k), (x_k, s_k + t - T_k)\right) \\ &= d^f\left((f^{j+N_k}(x), s_0 + g(t) - N_k - j), (f^j(x_k), s_k + t - T_k - j)\right) \\ &\le d^f\left((f^{j+N_k}(x), s_0 + g(t) - N_k - j), (f^{j+N_k}(x), s_k + t - T_k - j)\right) \\ &\quad + d^f\left((f^{j+N_k}(x), s_k + t - T_k - j), (f^j(x_k), s_k + t - T_k - j)\right) \end{aligned}$$

$$\begin{aligned}
&\leq |s_0 + g(t) - N_k - j - (s_k + t - T_k - j)| \\
&\quad + (s_k + t - T_k - j)d(f^{j+N_k+1}(x),\ f^{j+1}(x_k)) \\
&\quad + (1 - s_k - t + T_k + j)d(f^{j+N_k}(x),\ f^{j}(x_k)) \\
&< \delta' + \tfrac{1}{2}(1 - (s_k + t - T_k - j))\varepsilon + \tfrac{1}{2}(s_k + t - T_k - j)\varepsilon \\
&\leq \frac{\varepsilon}{2} + \frac{\varepsilon}{2} = \varepsilon.
\end{aligned}$$

Hence, ξ is ε-shadowed by (x, r), and therefore $(x,\ s_0) \in Sh(\phi^f)$. This completes the proof.

□

The following proposition characterizes shadowing flows on compact metric spaces in terms of shadowable points. More precisely, we show that a flow ϕ on a compact metric space X is shadowing if and only if $Sh(\phi) = X$.

Proposition 8.3 *Let ϕ be a flow on a compact metric space X. Then ϕ is shadowing if and only if $Sh(\phi) = X$.*

Proof Suppose that every point of X is shadowable for ϕ, and ϕ is not shadowing. Then there exists $\varepsilon > 0$ such that, for each $k \in \mathbb{N}$, there is a $(\frac{1}{k}, 1, 2)$-pseudo orbit $\xi^k = \{(x_n^k, t_n^k)\}_{n\in\mathbb{Z}}$ of ϕ which cannot be ε-shadowed. Since X is compact, we may assume that $x_0^k \to p$ and $t_0^k \to t_0$ as $k \to \infty$ for some $p \in X$ and $t_0 \in [1, 2]$. Without loss of generality, we may further assume that $t_0^k \leq t_0$ for all $k \in \mathbb{N}$.

Since p is shadowable for ϕ, choose $0 < \delta < \frac{\varepsilon}{3}$ such that every δ-pseudo orbit of ϕ through p is $\frac{\varepsilon}{3}$-shadowed by some point in X. By the continuity of ϕ, there exists $0 < \delta_0 < \frac{\delta}{2}$ such that if $d(x, y) < \delta_0$ and $|t - s| < \delta_0$ for $t, s \in [0, 2]$, then

$$d(\phi_t(x), \phi_s(y)) \leq \frac{\delta}{2}. \tag{8.3}$$

Let $k \in \mathbb{N}$ be such that

$$\frac{1}{k} < \delta_0, \quad d(x_0^k, p) < \delta_0, \quad \text{and} \quad |t_0 - t_0^k| < \delta_0. \tag{8.4}$$

We define a sequence $\xi = \{(y_n, s_n)\}$ by

$$(y_n, s_n) = \begin{cases} (p, t_0), & \text{if } n = 0, \\ (x_n^k, t_n^k), & \text{if } n \neq 0, \end{cases} \quad \text{for all } n \in \mathbb{Z}.$$

By (8.3) and (8.4), we first observe that

$$\begin{aligned} d(\phi_{s_0}(y_0), y_1) &= d(\phi_{t_0}(p), x_1^k) \\ &\le d(\phi_{t_0}(p), \phi_{t_0^k}(x_0^k)) + d(\phi_{t_0^k}(x_0^k), x_1^k) \\ &\le \frac{\delta}{2} + \frac{1}{k} < \delta, \end{aligned}$$

and similarly,

$$\begin{aligned} d(\phi_{s_{-1}}(y_{-1}), y_0) &= d(\phi_{t_{-1}^k}(x_{-1}^k), p) \\ &\le d(\phi_{t_{-1}^k}(x_{-1}^k), x_0^k) + d(x_0^k, p) \\ &\le \frac{1}{k} + \delta_0 < \delta. \end{aligned}$$

Hence, $\xi = \{(y_n, s_n)\}_{n\in\mathbb{Z}}$ is a $(\delta, 1, 2)$-pseudo orbit of ϕ with $y_0 = p$. Since p is shadowable for ϕ, there exist $z \in X$ and $h \in \mathcal{H}$ such that

$$d(\phi_{h(t)}(z), \phi_{t-T_n}(y_n)) < \frac{\varepsilon}{3}, \quad \text{for all } T_n \le t < T_{n+1} \text{ and } n \in \mathbb{Z},$$

where T_n is the nth sum associated to ξ.

We now show that ξ^k is ε-shadowed by z. In fact, let T_n^k denote the nth sum associated by ξ^k. Observe that $T_n^k = T_n$ for all $n \le 0$ and $T_n^k \le T_n$ for all $n > 0$. Let $t \in \mathbb{R}$. If $t < 0$, choose $n < 0$ such that $T_n^k \le t < T_{n+1}^k$. Then

$$d(\phi_{h(t)}(z), \phi_{t-T_n^k}(x_n^k)) = d(\phi_{h(t)}(z), \phi_{t-T_n}(y_n)) < \varepsilon.$$

If $t \ge 0$, choose $n \ge 0$ such that $T_n^k \le t < T_{n+1}^k$, and consider the following two cases.

Case 1 $T_n^k \le t < T_{n+1}$. Since $d(x_0^k, y_0) < \delta_0$ and $|T_n - T_n^k| = |t_0 - t_0^k| < \delta_0$, by (8.3), we have

$$\begin{aligned} d(\phi_{h(t)}(z), \phi_{t-T_n^k}(x_n^k)) &\le d(\phi_{h(t)}(z), \phi_{t-T_n}(y_n)) + d(\phi_{t-T_n}(y_n), \phi_{t-T_n^k}(x_n^k)) \\ &\le \frac{\varepsilon}{2} + \frac{\delta}{2} < \varepsilon. \end{aligned}$$

Case 2 $T_{n+1} \le t < T_{n+1}^k$. Since $d(\phi_{t_n^k}(x_n^k), x_{n+1}^k) < \delta_0$ and $|T_{n+1} - T_{n+1}^k| = |t_0 - t_0^k| < \delta_0$, by (8.3), we have

$$\begin{aligned} d(\phi_{h(t)}(z), \phi_{t-T_n^k}(x_n^k)) &\le d(\phi_{h(t)}(z), \phi_{t-T_{n+1}}(y_{n+1})) \\ &\quad + d(\phi_{t-T_{n+1}}(y_{n+1}), \phi_{t-T_{n+1}^k}(x_{n+1}^k)) \\ &\quad + d(\phi_{t-T_{n+1}^k}(x_{n+1}^k), \phi_{t-T_n^k}\phi_{t_n^k}(x_n^k)) \\ &\le \frac{\varepsilon}{3} + \frac{\delta}{2} + \frac{\delta}{2} < \varepsilon. \end{aligned}$$

Consequently,

$$d(\phi_{h(t)}(z), \phi_{t-T_n^k}(x_n^k)) < \varepsilon \quad \text{for all } t \in \mathbb{R}.$$

Hence ξ^k is ε-shadowed by z, which contradicts the choice of ξ^k. This contradiction completes the proof.

□

In Theorem 4.1, we provided a characterization of shadowing for homeomorphisms by using shadowing measures. In the following theorem, we prove a flow version of Theorem 4.1. For this, we need a lemma.

Lemma 8.3 *Let ϕ be a flow on a compact metric space X. Then:*

(1) $Sh_{\varepsilon,T}(\phi) \subset Sh_{2\varepsilon,S}(\phi)$ for any $T, S > 0$;
(2) $x \in Sh_{\varepsilon,T}(\phi)$ if and only if there is $\delta > 0$ such that any (δ, T)-pseudo orbit $\{(x_n, t_n)\}_{n\in\mathbb{Z}}$ through x with $T \le t_n \le 2T$ $(n \in \mathbb{Z})$ can be ε-shadowed.

Proof See Lemma 3.2 in [4].

□

Theorem 8.2 *Let ϕ be a flow on a compact metric space X. Then the followings are pairwise equivalent:*

(1) ϕ is shadowing;
(2) ϕ is measure shadowing;
(3) ϕ is invariant measure shadowing;
(4) $\mu(Sh(\phi)) = 1$ for all $\mu \in \mathcal{M}(X, \phi)$.

Proof (1) $\Rightarrow$ (2) $\Rightarrow$ (3) $\Rightarrow$ (4). It is clear from the definitions.

(4) $\Rightarrow$ (1). Suppose $\mu(Sh(\phi)) = 1$ for any $\mu \in \mathcal{M}(X, \phi)$. We first prove the following claim.

Claim: If $x \in Sh_\varepsilon(\phi)$ for some $\varepsilon > 0$, then for any $T > 0$ there is $\delta_T > 0$ such that every (δ_T, T)-pseudo orbit $\{(y_n, t_n)\}_{n\in\mathbb{Z}}$ with $d(y_0, x) < \delta_T$ can be 3ε-shadowed.

Let $\varepsilon > 0$ and $T > 0$ be arbitrary, and take $x \in Sh_\varepsilon(\phi)$. By Lemma 8.3, we have $x \in Sh_{2\varepsilon,T}(\phi)$. Therefore, there exists $\delta > 0$ such that every (δ, T)-pseudo orbit $\{(x_n, t_n)\}_{n\in\mathbb{Z}}$ through x with $T \le t_n \le 2T$ $(n \in \mathbb{Z})$ can be 2ε-shadowed. Choose $0 < \delta_T < \delta/2$ such that if $d(y, z) < \delta_T$ then $d(\phi_t(y), \phi_t(z)) < \min\{\delta/2, \varepsilon\}$ for all $t \in [0, 2T]$. Let $\{(y_n, t_n)\}_{n\in\mathbb{Z}}$ be a (δ_T, T)-pseudo orbit satisfying $d(y_0, x) < \delta_T$. Define a sequence $\{(z_n, s_n)\}_{n\in\mathbb{Z}}$ by

$$(z_n, s_n) = (x, t_0) \text{ for } n = 0, \text{ and } (z_n, s_n) = (y_n, t_n) \text{ for } n \neq 0.$$

Since

$$d(\phi_{t_{-1}}(y_{-1}), x) \leq d(\phi_{t_{-1}}(y_{-1}), y_0) + d(y_0, x) < \delta$$

and

$$d(\phi_{t_0}(x), y_1) \leq d(\phi_{t_0}(x), \phi_{t_0}(y_0)) + d(\phi_{t_0}(y_0), y_1) < \delta,$$

we see that $\{(z_n, s_n)\}_{n\in\mathbb{Z}}$ is a (δ, T)-pseudo orbit of ϕ through x. Since $x \in Sh_{2\varepsilon,T}(\phi)$, there is $z \in X$ and $h \in \mathcal{H}$ such that

$$d(\phi_{h(t)}(z), \phi_{t-T_n}(x_n)) \leq 2\varepsilon, \quad \forall t \in [T_n, T_{n+1}) \text{ with } n \in \mathbb{Z},$$

where T_n is given as before. For $T_0 \leq t < T_1$, we obtain

$$d(\phi_{h(t)}(z), \phi_t(y)) \leq d(\phi_{h(t)}(z), \phi_t(x)) + d(\phi_t(x), \phi_t(y)) \leq 3\varepsilon.$$

Thus, $\{(y_n, t_n)\}_{n\in\mathbb{Z}}$ is 3ε-shadowed by z, and so the claim is proved.

Let $x \in X$. Since the ω-limit set $\omega(x, \phi)$ is a nonempty, closed, invariant set, there exists a measure $\mu \in \mathcal{M}(X, \phi)$ such that $\mu(\omega(x, \phi)) = 1$. Because $\mu(Sh(\phi)) = 1$, it follows that $\omega(x, \phi) \cap Sh(\phi) \neq \emptyset$. Let $y \in \omega(x, \phi) \cap Sh(\phi)$. For any $\varepsilon > 0$, the previous claim ensures that there exists $\delta_1 > 0$ such that every $(\delta_1, 1)$-pseudo orbit $\{(y_n, t_n)\}_{n\in\mathbb{Z}}$ with $d(y, y_0) < \delta_1$ is ε-shadowed. Since $y \in \omega(x, \phi)$, there exists $T > 0$ such that $d(\phi_T(x), y) < \delta_1$. Hence, $\phi_T(x) \in Sh_\varepsilon(\phi)$.

Next, we show that $x \in Sh_{8\varepsilon}(\phi)$. By the previous claim, there exists $0 < \delta_T < \varepsilon$ such that every (δ_T, T)-pseudo orbit $\{(y_n, t_n)\}_{n\in\mathbb{Z}}$ with $d(y_0, \phi_T(x)) < \delta_T$ can be 3ε-shadowed. Choose $\delta' < \delta_T$ such that if $d(x, y) < \delta'$, then $d(\phi_t(x), \phi_t(y)) < \delta_T$ for all $t \in [-T, T]$. Let $\{(x_n, t_n)\}_{n\in\mathbb{Z}}$ be a (δ', T)-pseudo orbit of ϕ through x with $T \leq t_n \leq 2T$ for all $n \in \mathbb{Z}$. We now define a sequence $\{(y_n, s_n)\}_{n\in\mathbb{Z}}$ by

$$(y_n, s_n) = \begin{cases} (x_{n+1},\ t_{n+1}), & \text{if } n < -1, \\ (x_0,\ T), & \text{if } n = -1, \\ (\phi_{T-t_0}(x_1),\ t_1 + t_0 - T), & \text{if } n = 0, \\ (x_{n+1},\ t_{n+1}), & \text{if } n > 0. \end{cases}$$

Since $d(\phi_{t_0}(x_0), x_1) < \delta'$, we have

$$d(\phi_{T-t_0}(\phi_{t_0}(x_0)), \phi_{T-t_0}(x_1)) = d(\phi_T(x_0), \phi_{T-t_0}(x_1)) < \delta_T.$$

Hence, $\{(y_n, s_n)\}_{n\in\mathbb{Z}}$ is a (δ_T, T)-pseudo orbit of ϕ through $\phi_{T-t_0}(x_1)$. Moreover,

$$d(y_0, \phi_T(x)) = d(\phi_{T-t_0}(x_1), \phi_T(x_0)) < \delta_T.$$

Therefore, there exist $z \in X$ and $g \in \mathcal{H}$ such that

$$d(\phi_{g(t)}(z), \phi_{t-T_n}(y_n)) \leq 3\varepsilon, \quad \forall t \in [T_n, T_{n+1}) \text{ with } n \in \mathbb{Z},$$

where T_n is given as before. Furthermore,

$$d(\phi_{g(t)}(z), \phi_t(\phi_T(x))) \leq d(\phi_{g(t)}(z), \phi_t(\phi_{-t_0+T}(x_1))) + d(\phi_t(\phi_{-t_0+T}(x_1)), \phi_t(\phi_T(x))) < 4\varepsilon,$$

for all $t \in [0, t_0 - T]$. Thus, $\{(x_n, t_n)\}_{n\in\mathbb{Z}}$ is 4ε-shadowed by z. By Lemma 8.3, it follows that $x \in Sh_{8\varepsilon}(\phi)$. Since ε is arbitrary, we conclude that $x \in Sh(\phi)$, and hence $Sh(\phi) = X$. Finally, by Corollary 8.3, we derive that ϕ is shadowing on X.

□

8.2 Topologically Stable Measures for Flows

In this section, we prove a measurable version of stability theorem for flows. For this, we first introduce the notion of topologically stable measures for flows. Let ϕ be a flow on a compact metric space X.

Definition 8.6 A Borel measure $\mu \in \mathcal{M}(X)$ is *topologically stable* for ϕ if for every $\varepsilon > 0$ there exists $\delta > 0$ such that for any flow ψ on X with $d_{C^0}(\phi, \psi) < \delta$, there is an upper semicontinuous compact-valued map $H : \mathrm{Dom}(H) \subset X \to 2^X$ with ψ-invariant $\mathrm{Dom}(H)$ such that:

(1) $\mu(X \setminus \mathrm{Dom}(H)) = 0$,
(2) $\mu(H(x)) = 0$ for all $x \in \mathrm{Dom}(H)$,
(3) $d(H, id) \leq \varepsilon$,
(4) $H(\psi_{\mathbb{R}}(x)) = \phi_{\mathbb{R}}(H(x))$ for all $x \in \mathrm{Dom}(H)$.

In the following theorem, we present a measurable version of Walters' stability theorem for flows.

Theorem 8.3 *Let ϕ be a flow on a compact metric space X without fixed points, and $\mu \in \mathcal{M}(X)$. If μ is expansive and shadowing for ϕ, then μ is topologically stable for ϕ.*

Proof Suppose that μ is expansive and shadowing for ϕ. Let $e > 0$ be an expansive constant for μ. For $\varepsilon > 0$, we let $\varepsilon' = \min\{\frac{e}{2}, \varepsilon\}$. Since μ is shadowing for ϕ and $Fix(\phi) = \emptyset$, there are $\delta > 0$ and a Borel set B such that any $(\delta, 1)$-pseudo orbit of ϕ through B can be strongly ε'-shadowed by a point in X. Let ψ be a flow on X with $d_{C^0}(\phi, \psi) \leq \delta$. For each $x \in X$, we define

$$H(x) = \{y \in X : \exists h \in \mathcal{H}(\varepsilon') \text{ s.t. } d(\phi_{h(t)}(y), \psi_t(x)) \leq \varepsilon' \quad \forall t \in \mathbb{R}\}$$

and denote by $\mathrm{Dom}(H)$ the collection of $x \in X$ such that $H(x)$ is nonempty. We first observe that $\mathrm{Dom}(H)$ is nonempty. Indeed, let $x \in B$. Since

$$d(\phi_1(\phi_n(x)), \phi_{n+1}(x)) \leq \delta, \quad \forall n \in \mathbb{Z},$$

we get $\{\psi_n(x)\}_{n\in\mathbb{Z}}$ is a δ-pseudo orbit of ϕ through B. Since μ is shadowing for ϕ, there are $y \in X$ and $h \in \mathcal{H}(\varepsilon')$ such that

$$d(\phi_{h(t)}(y), \psi_t(x)) \leq \varepsilon', \quad \forall t \in \mathbb{R}.$$

Then $y \in H(x)$, and so $x \in \mathrm{Dom}(H)$. Consequently, we get $B \subset \mathrm{Dom}(H)$, and so $\mathrm{Dom}(H)$ is nonempty.

Moreover, we see that $H(x)$ is compact for all $x \in \mathrm{Dom}(H)$. In fact, let $x \in \mathrm{Dom}(H)$ and $\{y_i\}_{i\in\mathbb{N}}$ be a sequence in $H(x)$ such that $y_i \to y$ for some $y \in X$. Since $y_i \in H(x)$, there is a homeomorphism $h_i \in \mathcal{H}(\varepsilon')$ such that

$$d(\phi_{h_i(t)}(y_i), \psi_t(x)) \leq \varepsilon', \quad \forall t \in \mathbb{R}.$$

Note that $\{h_i\}_{i\in\mathbb{N}}$ is pointwisely bounded and equicontinuous. By the Arzela–Ascoli theorem, there are subsequence of $\{h_i\}_{i\in\mathbb{N}}$, we still denote it by $\{h_i\}_{i\in\mathbb{N}}$, such that $\{h_i\}_{i\in\mathbb{N}}$ converges compactly to h for some $h \in \mathcal{H}(\varepsilon')$. For any fixed $t \in \mathbb{R}$, we take $N \in \mathbb{N}$ such that $t \in [-N, N]$. We have

$$d(\phi_{h(t)}(y), \psi_t(x)) = \lim_{i\to\infty} d(\phi_{h_i(t)}(y_i), \psi_t(x)) \leq \varepsilon'.$$

Then we get $y \in H(x)$, and so $H(x)$ is compact. Hence we can define a compact-valued map $H : \mathrm{Dom}(H) \to 2^X$.

Next, we show that $\mathrm{Dom}(H)$ is ψ-invariant. Let $s \in \mathbb{R}$ and $y \in H(x)$. Then there is $h \in \mathcal{H}(\varepsilon')$ such that

$$d(\phi_{h(t)}(y), \psi_t(x)) \leq \varepsilon', \quad \forall t \in \mathbb{R}.$$

Define a homeomorphism $g : \mathbb{R} \to \mathbb{R}$ by $g(t) = h(t+s) - h(s)$ for all $t \in \mathbb{R}$. It is clear that $g \in \mathcal{H}(\varepsilon')$. Since

$$d(\phi_{g(t)}\phi_{h(s)}(y), \psi_t\psi_s(x)) = d(\phi_{h(t+s)}(y), \psi_{t+s}(x)) \leq \varepsilon', \quad \forall t \in \mathbb{R},$$

we get $\phi_{h(s)}(y) \in H(\psi_s(x))$, and so $\psi_s(x) \in \mathrm{Dom}(H)$.

(1) $\mathrm{Dom}(H)$ is measurable, and $\mu(X \setminus \mathrm{Dom}(H)) = 0$. We observe that $\mathrm{Dom}(H)$ is closed in X. Let $\{x_n\}_{n\in\mathbb{N}}$ be a sequence in $\mathrm{Dom}(H)$ such that $x_n \to x$ for some $x \in X$. For each $n \in \mathbb{N}$, since $x_n \in \mathrm{Dom}(H)$, there are $y_n \in X$ and $h_n \in \mathcal{H}(\varepsilon')$ such that

$$d(\phi_{h_n(t)}(y_n), \psi_t(x_n)) \leq \varepsilon', \quad \forall t \in \mathbb{R}.$$

Since X is compact and $\{h_n\}_{n\in\mathbb{N}}$ is pointwisely bounded and equicontinuous, there is a subsequence $\{n_k\}_{k\in\mathbb{N}}$ such that $y_{n_k} \to y$ and h_{n_k} compactly converges to h for some $y \in X$ and $h \in \mathcal{H}(\varepsilon')$. For any $t \in \mathbb{R}$, there is $N > 0$ such that $t \in [-N, N]$ and $h_{n_k}|_{[-N,N]} \to h|_{[-N,N]}$ uniformly. Then we get

$$d(\phi_{h(t)}(y), \psi_t(x)) = \lim_{k\to\infty} d(\phi_{h_{n_k}(t)}(y_{n_k}), \psi_t(x)) \le \varepsilon'.$$

Then $y \in H(x)$, and so $x \in \mathrm{Dom}(H)$. Hence we get $\mathrm{Dom}(H)$ is closed. Consequently, we obtain that $\mu(X \setminus \mathrm{Dom}(H)) = 0$.

(2) $\mu(H(x)) = 0$ for all $x \in \mathrm{Dom}(H)$. Let $x \in \mathrm{Dom}(H)$ and $y \in H(x)$. Then there is $h_y \in \mathcal{H}(\varepsilon')$ such that

$$d(\phi_{h_y(t)}(y), \psi_t(x)) \le \varepsilon'. \quad \forall t \in \mathbb{R}.$$

For any $z \in H(x)$, there is $h_z \in \mathcal{H}(\varepsilon')$ such that

$$d(\phi_{h_z(t)}(z), \psi_t(x)) \le \varepsilon', \quad \forall t \in \mathbb{R}.$$

Then

$$\begin{aligned} d(\phi_t(y), \phi_{h_z\circ h_y^{-1}(t)}(z)) &\le d(\phi_t(y), \phi_{h_y^{-1}}(x)) + d(\phi_{h_y^{-1}}(x), \phi_{h_z\circ h_y^{-1}(t)}(z)) \\ &\le 2\varepsilon' < e. \end{aligned}$$

Then $z \in \Gamma_e^{\phi}(y)$, and so $H(x) \subset \Gamma_e^{\phi}(y)$. Since f is μ-expansive, we get $\mu(H(x)) \le \mu(\Gamma_e^{\phi}(y)) = 0$.

(3) $d(H, id) \le \varepsilon$. It is clear by the definition of H.

(4) H is upper semicontinuous. Suppose that H is not upper semicontinuous at $x \in \mathrm{Dom}(H)$. Then there are a constant $\varepsilon_0 > 0$ and a sequence $\{y_n\}_{n\in\mathbb{N}}$ with $y_n \to x$ and $H(y_n) \setminus B(H(x), \varepsilon_0) \neq \emptyset$. For each $n \in \mathbb{N}$, let $z_n \in H(y_n) \setminus B(H(x), \varepsilon_0)$. Then there is $h_n \in \mathcal{H}(\varepsilon')$ such that

$$d(\phi_{h_n(t)}(z_n), \psi_t(y_n)) \le \varepsilon', \quad \forall t \in \mathbb{R}.$$

since X is compact and $\{h_n\}_{n\in\mathbb{N}}$ is pointwisely bounded and equicontinuous, there is a subsequence $\{n_k\}_{k\in\mathbb{N}}$ such that $z_{n_k} \to z$ and h_{n_k} compactly converges to h for some $z \in X$ and $h \in \mathcal{H}(\varepsilon')$. Note that $z \notin B(H(x), \varepsilon_0)$.

On the other hand, for any $t \in \mathbb{R}$, there is $N > 0$ such that $t \in [-N, N]$ and $h_{n_k}|_{[-N,N]} \to h|_{[-N,N]}$ uniformly. Then we have

$$d(\phi_h(t)(z), \psi_t(x)) = \lim_{k\to\infty} d(\phi_{h_{n_k}(t)}(z_{n_k}), \psi_t(y_{n_k})) \le \varepsilon'.$$

It implies that $z \in H(x)$. The contradiction shows that H is upper semicontinuous.

(5) $H(\psi_{\mathbb{R}}(x)) = \phi_{\mathbb{R}}(H(x))$ for all $x \in \mathrm{Dom}(H)$.

Let $x \in \mathrm{Dom}(H)$ and $y \in H(\psi_s(x))$ for some $s \in \mathbb{R}$. Then there is $h \in \mathcal{H}(\varepsilon')$ such that

$$d(\phi_{h(t)}(y), \psi_{t+s}(x)) \le \varepsilon', \quad \forall t \in \mathbb{R}.$$

Let $\tau = t + s$. We have

$$d(\phi_{h(\tau-s)-h(-s)}\phi_{h(-s)}(y), \psi_\tau(x)) \le \varepsilon', \quad \forall \tau \in \mathbb{R}.$$

Then $\phi_{h(-s)}(y) \in H(x)$, and so $y \in \phi_{-h(-s)}(H(x)) \subset \phi_{\mathbb{R}}(H(x))$.

Conversely, let $y \in H(x)$ and $s \in \mathbb{R}$. We claim that $\phi_s(y) \in H(\psi_{\mathbb{R}}(x))$. Since $y \in H(x)$, there is $h \in \mathcal{H}(\varepsilon')$ such that

$$d(\phi_{h(t)}(y), \psi_t(x)) \le \varepsilon', \quad \forall t \in \mathbb{R}.$$

Let $\tau = h^{-1}(s)$ and define a homeomorphism $g : \mathbb{R} \to \mathbb{R}$ by $g(t) = h(t + \tau) - h(\tau)$ for all $t \in \mathbb{R}$. It is clear that $g \in \mathcal{H}(\varepsilon')$. Since

$$d(\phi_{g(t)}\phi_s(y), \psi_t\psi_\tau(x)) = d(\phi_{h(t+\tau)-h(\tau)}\phi_{h(\tau)}(y), \psi_{t+\tau}(x)) \le \varepsilon', \quad \forall t \in \mathbb{R},$$

we get $\phi_s(y) \in H(\psi_\tau(x)) \subset H(\psi_{\mathbb{R}}(x))$. This completes the proof. □

Chapter 9
Measurable Spectral Decomposition for Flows

The purpose of this chapter is to present a measurable version of the spectral decomposition theorem for flows. The contents of this chapter are based on [28]. In Sect. 9.1, we introduce another type of expansive measure (called a geometric expansive measure) for flows on a compact metric space. This notion is used to characterize the geometric expansiveness for flows. In Sect. 9.2, we introduce the notion of invariant measure geometric expansive, and prove an extension theorem for invariant measure geometric expansive flows on its chain recurrent set. In Sect. 9.3, we present a measurable version of spectral decomposition theorem which says that if a flow ϕ is invariant measure 1-expansive and shadowing on its chain recurrent set $CR(\phi)$ then the flow ϕ admits the spectral decomposition.

9.1 Geometric Expansive Measures for Flows

Throughout the section, we assume ϕ is a flow on a compact metric space X. We now introduce another type of expansive measure, called a geometric expansive measure for flows.

Definition 9.1 A Borel measure $\mu \in \mathcal{M}(X)$ is said to be *geometric N-expansive* ($N \in \mathbb{N}$) for ϕ if there is $\delta > 0$ (called an *geometric N-expansive constant* of μ) such that for every $x \in X$ there is a subset $F \subset X$ containing x with the cardinality $card(F) \leq N$, and $\mu(\Gamma_\delta^\phi(x) \setminus \bigcup_{y \in F} \phi_{\mathbb{R}}(y)) = 0$.

It is clear that if a Borel measure $\mu \in \mathcal{M}(X)$ is ϕ-orbit vanishing, then μ is expansive for ϕ if and only if it is geometric expansive for ϕ. However, if μ is not ϕ-orbit vanishing, then the two notions are not equivalent in general. In fact, every Dirac measure δ_x ($x \in X$) on X is geometric expansive for any flow ϕ on X, but it is not expansive for ϕ.

K. Lee et al., *Topological Dynamics from a Measure-Theoretic Viewpoint*, Springer Asia Pacific Mathematics Series 14, https://doi.org/10.1007/978-981-95-6390-6_9

Definition 9.2 A flow ϕ is said to be *measure geometric* N*-expansive* (resp. *invariant measure geometric* N*-expansive*) if every Borel measure (resp. invariant Borel measure) is geometric N-expansive for ϕ.

In the following proposition, we will show that if ϕ is invariant measure N-expansive, then the N-expansive constant can be chosen uniformly for every $\mu \in \mathcal{M}(X, \phi)$.

Proposition 9.1 *Let ϕ be a flow on a compact metric space X. Then a flow ϕ is invariant measure N-expansive if and only if there is a constant $\delta > 0$ such that for every $\mu \in \mathcal{M}(X, \phi)$ and $x \in X$, there is a subset $F \subset X$ containing x with the cardinality $card(F) \leq N$ such that*

$$\mu(\Gamma_\delta^\phi(x) \setminus \bigcup_{y \in F} \phi_{\mathbb{R}}(y)) = 0.$$

Proof We only need to prove the sufficiency direction. Suppose, on the contrary, that for each $n \in \mathbb{N}$, there exist $x_n \in X$ and $\mu_n \in \mathcal{M}(X, \phi)$ such that for any set $F \subset X$ containing x_n with the cardinality $card(F) \leq N$, there is a Borel set $B_F \subset \Gamma_{1/n}^\phi(x_n)$ such that

$$\mu(B_F \setminus \bigcup_{y \in F} \phi_{\mathbb{R}}(y)) > 0.$$

Define a Borel measure μ on X by

$$\mu(A) = \sum_{n=1}^{\infty} \frac{\mu_n(A)}{2^n}$$

for any Borel subset A of X. It is clear that μ is an invariant measure. Since ϕ is measure N-expansive, μ is N-expansive. Let $\delta > 0$ be an N-expansive constant of μ, and choose a natural number m with $1/m < \delta$. Let F be a subset of X containing x_m and with the cardinality $card(F) \leq N$. By the above assumption, there is a Borel set $B_F \subset \Gamma_{1/m}^\phi(x_m)$ such that

$$\mu\left(B_F \setminus \bigcup_{y \in F} \phi_{\mathbb{R}}(y)\right) \geq \frac{\mu_m\left(B_F \setminus \bigcup_{y \in F} \phi_{\mathbb{R}}(y)\right)}{2^m} > 0.$$

Since F is arbitrary, we get a contradiction. This completes the proof. □

By Theorem 7.1(1), we know that if a Borel measure $\mu \in \mathcal{M}(X)$ is expansive for a flow ϕ, then supp$(\mu) \cap Fix(\phi) = \emptyset$. This implies that every measure expansive flow on a compact smooth manifold has no fixed points. However, an invariant measure geometric expansive flow can admit a fixed point as we see in the following example.

Example 9.1 Let S^2 be the unit sphere in $\mathbb{R}^3$, and let ϕ be the flow generated by the vector field F on S^2 defined by

$$F(p) = (x^2 + y^2)(xz, yz, -x^2 - y^2) \text{ for } p = (x, y, z) \in S^2.$$

We observe that $Fix(\phi) = \Omega(\phi) = \{N = (0, 0, 1), S = (0, 0, -1)\}$, the ω-limit set $\omega(p, \phi)$ and α-limit set $\alpha(p, \phi)$ of $p \in S^2 \setminus Fix(\phi)$ are given by

$$\omega(p, \phi) = \{S\} \text{ and } \alpha(p, \phi) = \{N\}.$$

Since $\text{supp}(\mu) \subset \Omega(\phi)$ for any $\mu \in \mathcal{M}(S^2, \phi)$, we have

$$\mathcal{M}(S^2, \phi) = \{\lambda\delta_N + (1 - \lambda)\delta_S \mid \lambda \in [0, 1]\},$$

where δ_p is the Dirac measure centered at p. For any $p \in S^2$ and $\mu \in \mathcal{M}(S^2, \phi)$, it is easy to see that $\mu(\Gamma^{\phi}_{1/4}(p) \setminus \phi_{\mathbb{R}}(p)) = 0$. Then ϕ is invariant measure geometric expansive.

Theorem 9.1 *Let ϕ be a flow on a compact metric space X. A flow ϕ is measure geometric N-expansive if and only if it is geometric N-expansive.*

Proof Suppose a flow ϕ is measure geometric N-expansive, but not geometric N-expansive. Then for each $n \in \mathbb{N}$, there exist $x_n \in X$ and a finite sequence $\{y_n^1, \ldots, y_n^{N+1}\}$ such that:

- $y_n^i \in \Gamma^{\phi}_{1/n}(x_n)$ for all $1 \leq i \leq N + 1$,
- $\phi_{\mathbb{R}}(y_n^i) \cap \phi_{\mathbb{R}}(y_n^j) = \emptyset$ for $i \neq j \in \{1, \ldots, N + 1\}$.

Define a Borel measure μ on X by

$$\mu(A) = \sum_{n=1}^{\infty} \sum_{i=1}^{N+1} \frac{\delta_{y_n^i}(A)}{2^n(N + 1)}$$

for any Borel subset A of X. Since ϕ is measure geometric N-expansive, μ is geometric N-expansive. Let $\delta > 0$ be a geometric N-expansive constant of μ, and choose $m \in \mathbb{N}$ with $\frac{1}{m} < \delta$. Let $F \subset X$ be such that $x \in F$ and the cardinality $card(F) \leq N$. Then there exists $i \in \{1, \ldots, N + 1\}$ such that $y_m^i \in \Gamma^{\phi}_{1/m}(x_m) \setminus F$. Hence we have

$$\mu\left(\Gamma^{\phi}_{\delta}(x_m) \setminus \bigcup_{y \in F} \phi_{\mathbb{R}}(y)\right) \geq \frac{\delta_{y_m^i}(\{y_m^i\})}{2^m} > 0.$$

The contradiction shows that ϕ is geometric expansive. The converse is clear, and this completes the proof.

□

By definition, we know that if a flow ϕ is N-expansive then it is $(N+1)$-expansive for any $N \in \mathbb{N}$. However, we note that the converse is not true in general.

For any Borel measure μ on X, there exists a suspending measure $T^f(\mu)$ on X^f defined implicitly by

$$\int_{X^f} h(y)dT^f(\mu)(\overline{x}) = \int_X \int_0^1 h(\phi_t(x))dtd\mu(x)$$

for every continuous map $h : X^f \to \mathbb{R}$. In Theorem 7.2, we proved that a Borel measure μ is expansive for f if and only if its suspending measure $T^f(\mu)$ is expansive for the suspension flow ϕ.

Definition 9.3 Let f be a homeomorphism on a compact metric space X. A Borel measure $\mu \in \mathcal{M}(X)$ is said to be *geometric expansive* for f if there is $\delta > 0$ such that $\mu(\Gamma_\delta^f(x) \setminus \mathcal{O}_f(x)) = 0$ for all $x \in X$. Here δ is called an *geometric expansive constant* of μ.

We can easily see that if a Borel measure μ is strongly expansive for f then it is geometric expansive for f: more precisely, if $\mu(\Gamma_\delta^f(x)) = \mu(\{x\})$ for all $x \in X$, then $\mu(\Gamma_\delta^f(x) \setminus \mathcal{O}_f(x)) = 0$ for all $x \in X$. However, the converse is not true in general as we can see in Example 6.1.

Theorem 9.2 *Let ϕ be the suspension flow of a homeomorphism f on a compact metric space X. Then a Borel measure μ on X is geometric expansive for f if and only if its suspending measure $T^f(\mu)$ on X^f is geometric expansive for ϕ.*

Proof The proof is similar to that of Theorem 7.2.

□

Note that every measure geometric expansive flow is measure expansive, but the converse is not true in general. In fact, we can construct a countably expansive flow ϕ which is not geometric expansive. Then the flow ϕ is measure expansive by Theorem 7.4, and it is not measure geometric expansive by Theorem 9.1. Moreover, we can construct a measure expansive flow which is not invariant measure geometric expansive.

Example 9.2 Let Σ_2 denote the space of sequences over two symbols 0 and 1, equipped with the metric

$$d_0(x, y) = \begin{cases} \frac{1}{2^n} \text{ if } & n = \max\{k \in \mathbb{N} | \ x_i = y_i \text{ for all } |i| < k\}, \\ 0 & \text{if } \ x = y, \end{cases}$$

for $x = (x_i)_{i\in\mathbb{Z}}$, $y = (y_i)_{i\in\mathbb{Z}} \in \Sigma_2$. Define the shift map $\sigma : \Sigma_2 \to \Sigma_2$ by

$$(\sigma(x))_i = x_{i+1}, \quad \forall x = (x)_{i\in\mathbb{Z}} \in \Sigma_2.$$

Then σ is an expansive shadowing homeomorphism on Σ_2.

For each $n \in \mathbb{N}$, choose a periodic point $p_n \in \Sigma_2$ of period n, and define

$$E = \bigcup_{n\in\mathbb{N}} \mathcal{O}_\sigma(p_n).$$

Take a disjoint copy F of E such that $\Sigma_2 \cap F = \emptyset$, and let $X = \Sigma_2 \cup F$. The set F can be enumerated by a bijection $b : E \to F$ that assigns to each element $\sigma^k(p_n)$ of E the corresponding point $b(\sigma^k(p_n)) := q_{nk}$ in F, for $n \in \mathbb{N}$ and $0 \le k < n$. Define a metric D on X by

$$D(x, y) = \begin{cases} d_0(x, y) & \text{if } x, y \in \Sigma_2, \\ \frac{1}{n} + d_0(x, \sigma^k(p_n)) & \text{if } x \in \Sigma_2,\ y = q_{nk} \in F, \\ \frac{1}{n} + \frac{1}{m} + d_0(\sigma^k(p_n), \sigma^l(p_m)) & \text{if } x = q_{nk},\ y = q_{ml} \in F. \end{cases}$$

Define a homeomorphism $f : X \to X$ by

$$f(x) = \begin{cases} \sigma(x) & \text{if } x \in \Sigma_2, \\ q_{n(k+1)} & \text{if } x = q_{nk},\ 0 \le k < n-1, \\ q_{n0} & \text{if } x = q_{n(n-1)}. \end{cases}$$

By Theorem 3.7, the map f is 2-expansive. Hence, its suspension flow ϕ under the constant height function $\tau = 1$ on X^f is measure expansive by Theorem 7.4.

On the other hand, for any $\delta > 0$, choose a sufficiently large $n \in \mathbb{N}$ and a small constant $\alpha > 0$ such that

$$\phi_{(-\alpha,\alpha)}((q_{n0}, 0)) \subset \Gamma_\delta^\phi((p_n, 0)).$$

Let $\mathcal{L}_{q_{n0}}$ denote the invariant normalized Lebesgue measure supported on the periodic orbit $\phi_{\mathbb{R}}(q_{n0}, 0)$ of ϕ in X^f. Then we have

$$\mathcal{L}_{q_{n0}}(\Gamma_\delta^\phi((p_n, 0)) \setminus \phi_{\mathbb{R}}((p_n, 0))) > \mathcal{L}_{q_{n0}}(\phi_{(-\alpha,\alpha)}((q_{n0}, 0))) > 0.$$

Therefore, ϕ is not invariant measure geometric expansive.

9.2 Extension of Geometric Expansive Measures

It is clear that if a flow ϕ is expansive on X, then it is also expansive on every ϕ-invariant set $\Lambda \subset X$. However, the converse is not true in general. For any ϕ-invariant set $\Lambda \subset X$, we say that ϕ is (invariant) measure geometric expansive on Λ if $\phi|_\Lambda$ is (invariant) measure geometric expansive.

In this section, we show that if a flow ϕ is invariant measure geometric expansive on its chain recurrent set $CR(\phi)$, then it is also invariant measure geometric expansive on the phase space X. However, we see that the invariant geometric expansive measures on the nonwandering set $\Omega(\phi)$ cannot be extended to the invariant geometric expansive measures on X as we can see in Example 9.3.

We now show that the invariant geometric expansive measures on the chain recurrent set $CR(\phi)$ can be extended to the invariant geometric expansive measures on X.

Theorem 9.3 *Let ϕ be a flow on a compact metric space X. Then ϕ is invariant measure geometric expansive on X if and only if it is invariant measure geometric expansive on its chain recurrent set $CR(\phi)$.*

To prove Theorem 9.3, we need some notations and two lemmas. We first use the following lemma which is essentially contained in [11]. We include its proof for the sake of completeness.

Lemma 9.1 *Let ϕ be a flow on a compact metric space X without fixed points. Then for every $\alpha > 0$ there is $\delta > 0$ such that for every $x, y \in X$ satisfying $y \in \Gamma_\delta^\phi(x)$ there is an increasing homeomorphism $h \in \mathcal{H}$ satisfying $d(\phi_t(x), \phi_{h(t)}(y)) < \alpha$, $\forall t \in \mathbb{R}$.*

Proof Let $T_0 > 0$ be as in Lemma 6.4. We first claim that for every $0 < T < \frac{T_0}{3}$, there exist $\delta_T > 0$ and $\tau_T > 0$ such that if $c \in \mathcal{C}$ satisfies $d(\phi_t(x), \phi_{c(t)}(y)) < \delta_T$ for all $t \in \mathbb{R}$ and for some $x, y \in X$, then $c(t+T) - c(t) \geq \tau_T$ for all $t \in \mathbb{R}$. In fact, for $0 < T < \frac{T_0}{3}$, we let $\gamma_T > 0$ with respect to T by Lemma 6.4, and $\delta_T < \frac{\gamma_T}{2}$. Then for any $t \in \mathbb{R}$ we have

$$
\begin{aligned}
& d(\phi_{c(t)}(y), \phi_{c(t+T)}(y)) \\
& \geq d(\phi_t(x), \phi_{t+T}(x)) - d(\phi_t(x), \phi_{c(t)}(y)) - d(\phi_{c(t+T)}(y), \phi_{t+T}(x)) \\
& \geq \gamma_T - 2\delta_T > 0.
\end{aligned}
$$

By the continuity of ϕ, there is $\tau_T > 0$ such that

$$
|c(t+T) - c(t)| \geq \tau_T, \quad \forall t \in \mathbb{R}.
$$

It remains to show that $c(T) > 0$. Suppose, on the contrary, that there is $0 < T < T_0/3$ such that for every $n \in \mathbb{N}$, there exist $x_n, y_n \in X$ and a continuous map $c_n \in \mathcal{C}$ such that

$$
d(\phi_t(x_n), \phi_{c_n(t)}(y_n)) < \frac{1}{n}, \quad \forall t \in \mathbb{R}
$$

and $c_n(T) < 0$. Since X is compact, we may assume $x_n \to x$, and so $y_n \to x$ for some $x \in X$. We now consider the following two cases.

Case 1 there is a subsequence $\{n_k\}_{k\in\mathbb{N}}$ such that $c_{n_k}(T) \geq T$ for all $k \in \mathbb{N}$. We assume that $c_{n_k}(T) \to -s$ for some $s \in [0, T]$. Then we have

$$
d(\phi_T(x), \phi_{-s}(x)) = \lim_{k\to\infty} d(\phi_T(x_{n_k}), \phi_{c_{n_k}(T)}(y_{n_k})) = 0.
$$

It implies that $\phi_{T+s}(x) = x$, which contradicts the choice of T_0.

Case 2 there is a subsequence $\{n_k\}_{k\in\mathbb{N}}$ such that $c_{n_k}(T) < -T$ for all $k \in \mathbb{N}$. For each $k \in \mathbb{N}$, take $t_{n_k} \in [0, T]$ such that $c_{n_k}(t_{n_k}) = -T$. We assume that $t_{n_k} \to s$ for some $s \in [0, T]$. Then we have

$$d(\phi_{-T}(x), \phi_s(x)) = \lim_{k\to\infty} d(\phi_{t_{n_k}}(x_{n_k}), \phi_{c_{n_k}(t_{n_k})}(y_{n_k})) = 0.$$

It implies that $\phi_{T+s}(x) = x$, which contradicts the choice of T_0. Consequently, we complete the proof of the claim.

Fix $\alpha > 0$ and take $0 < T < \frac{T_0}{3}$ such that

$$\sup_{(z,u)\in X\times[0,T]} d(z, \phi_u(z)) \le \frac{\alpha}{2}.$$

For this T we fix δ_T and τ_T as above and also $0 < \delta < \min\left(\delta_T, \frac{\alpha}{2}\right)$.

Now let $x, y \in X$ be such that $y \in \Gamma_\delta^\phi(x)$, i.e., there is $c \in \mathcal{C}$ such that

$$d(\phi_t(x), \phi_{c(t)}(y)) < \delta, \quad \forall t \in \mathbb{R}.$$

Since $\delta < \delta_T$, we have $d(\phi_t(x), \phi_{c(t)}(y)) < \delta_T$ for all $t \in \mathbb{R}$, and so, by the choice of δ_T, we get $c(t + T) - c(t) \ge \tau_T$ for every $t \in \mathbb{R}$. Define a continuous map $h : \mathbb{R} \to \mathbb{R}$ by $h(nT) = h(nT)$ for every $n \in \mathbb{Z}$, and extend it linearly on each interval $[nT, (n + 1)T]$. It is clear that $h(0) = 0$, and by the choice of τ_T we see that h is an increasing homeomorphism. Moreover, for $t \in \mathbb{R}$, there is $n \in \mathbb{Z}$ such that $t \in [nT, (n + 1)T]$. Take $s \in [nT, (n + 1)T]$ such that $h(t) = c(s)$. Then we have

$$\begin{aligned} d(\phi_t(x), \phi_{h(t)}(y)) &= d(\phi_t(x), \phi_{c(s)}(y)) \\ &\le d(\phi_t(x), \phi_s(x)) + d(\phi_s(x), \phi_{c(s)}(y)) \\ &\le \left(\sup_{(z,u)\in X\times[0,T]} d(z, \phi_u(z))\right) + \delta < \frac{\alpha}{2} + \frac{\alpha}{2} = \alpha. \end{aligned}$$

Hence $d(\phi_t(x), \phi_{h(t)}(y)) < \alpha$ for all $t \in \mathbb{R}$. This completes the proof. □

For any $\delta > 0$ and $x \in X$, we define

$$\Gamma_\delta^\phi(x, \mathcal{H}) = \{y \in X | d(\phi_t(x), \phi_{h(t)}(y)) \le \delta \text{ for some } h \in \mathcal{H} \text{ and all } t \in \mathbb{R}\}.$$

Lemma 9.2 *Let ϕ be a flow on a compact metric space X without fixed points. Then for any $\varepsilon > 0$, there is $\delta > 0$ such that*

$$\overline{\Gamma_\delta^\phi(x, \mathcal{H})} \subset \Gamma_\varepsilon^\phi(x, \mathcal{H}) \text{ for all } x \in X.$$

Proof Let $\varepsilon > 0$. By Lemma 9.1, there is $\gamma > 0$ such that $\Gamma_\gamma^\phi(x) \subset \Gamma_\varepsilon^\phi(x, \mathcal{H})$ for all $x \in X$. Moreover, by Lemma 7.8, there is $\delta > 0$ such that $\tilde{\Gamma}_\delta^\phi(x)$ is compact and $\Gamma_\delta^\phi(x) \subset \tilde{\Gamma}_\delta^\phi(x) \subset \Gamma_\gamma^\phi(x)$ for all $x \in X$. Consequently, for any $x \in X$, we get

$$\overline{\Gamma_\delta^\phi(x, \mathcal{H})} \subset \tilde{\Gamma}_\delta^\phi(x) \subset \Gamma_\gamma^\phi(x) \subset \Gamma_\varepsilon^\phi(x, \mathcal{H}).$$

This completes the proof. □

Proof of Theorem 9.3 Suppose that ϕ is invariant measure geometric expansive on its chain recurrent set $CR(\phi)$. Then there exists $\delta > 0$ such that, for every $x \in X$ and every invariant measure $\mu \in \mathcal{M}(CR(\phi), \phi)$,

$$\mu((\Gamma_{2\delta}^\phi(x) \cap CR(\phi)) \setminus \phi_{\mathbb{R}}(x)) = 0.$$

Claim 1: We show $\mu(\Gamma_{2\delta}^\phi(x) \setminus \phi_{\mathbb{R}}(x)) = 0$ for all $x \in CR(\phi)$ and $\mu \in \mathcal{M}(X, \phi)$.

Let $x \in CR(\phi)$, and let $B \subset X$ be a Borel set such that $B \subset \Gamma_{2\delta}^\phi(x) \setminus \phi_{\mathbb{R}}(x)$. For any $\mu \in \mathcal{M}(X, \phi)$, we define a measure $\nu \in \mathcal{M}(CR(\phi), \phi)$ by $\nu(A) = \mu(A)$ for a Borel set $A \subset CR(\phi)$. Since ϕ is invariant measure geometric expansive on $CR(\phi)$, it follows that $\mu(B \cap CR(\phi)) = \nu(B \cap CR(\phi)) = 0$. Because μ is invariant, we have $\mu(CR(\phi)^c) = 0$. Therefore,

$$\mu(B) = \mu(B \cap CR(\phi)) + \mu(B \cap CR(\phi)^c) = \nu(B \cap CR(\phi)) = 0.$$

Hence, $\mu(\Gamma_{2\delta}^\phi(x) \setminus \phi_{\mathbb{R}}(x)) = 0$, which completes the proof of Claim 1.

Let $Y = \{x \in X \mid \phi_{\mathbb{R}}(x) \subset X \setminus U_\delta(Fix(\phi))\}$, where $U_\delta(Fix(\phi))$ denotes the δ-neighborhood of $Fix(\phi)$. Clearly, Y is a closed and invariant subset of X. Let ψ be the restriction of ϕ to Y, that is, $\psi(x, t) = \phi(x, t)$ for any $x \in Y$ and $t \in \mathbb{R}$. Without loss of generality, we may assume that $Fix(\phi)$ and Y are nonempty. It is straightforward that $CR(\psi) \subset CR(\phi)$ and $Fix(\psi) = \emptyset$. We observe that ψ is invariant measure geometric expansive on its chain recurrent set $CR(\psi)$. To extend the invariant measures of ψ to those of ϕ, we define a map $i : \mathcal{M}(Y, \psi) \to \mathcal{M}(X, \phi)$ by

$$i(\mu)(A) = \mu(A \cap Y), \quad \forall \mu \in \mathcal{M}(Y, \psi),$$

and for every A is a Borel subset of X.

Claim 2: We show that ψ is invariant measure geometric expansive on Y.

More precisely, we claim that there exists $\varepsilon > 0$ such that $\mu(\Gamma_\varepsilon^\psi(x) \setminus \psi_{\mathbb{R}}(x)) = 0$ for all $x \in Y$ and $\mu \in \mathcal{M}(Y, \psi)$. Since ψ has no fixed points, Lemma 9.2 ensures that we can choose $0 < \gamma < \delta$ satisfying

$$\overline{\Gamma_{2\gamma}^\psi(x, \mathcal{H})} \subset \Gamma_\delta^\psi(x, \mathcal{H}) \subset \Gamma_\delta^\phi(x), \quad \forall x \in Y.$$

We first show that $\mu(\Gamma^{\psi}_{\gamma}(x, \mathcal{H}) \setminus \psi_{\mathbb{R}}(x)) = 0$ for all $x \in Y$ and $\mu \in \mathcal{M}(Y, \psi)$. Assume to the contrary that there exist an invariant measure $\mu \in \mathcal{M}(Y, \psi)$ and a point $x \in Y$ such that $\mu(B \setminus \psi_{\mathbb{R}}(x)) > 0$ for some Borel set $B \subset \Gamma^{\psi}_{\gamma}(x, \mathcal{H})$. By Claim 1, this implies $x \notin CR(\psi)$. Since $\mu(CR(\psi)) = 1$, there exists a point $p \in (B \setminus \psi_{\mathbb{R}}(x)) \cap CR(\psi)$. Then

$$B \subset \Gamma^{\psi}_{\gamma}(x, \mathcal{H}) \subset \Gamma^{\psi}_{2\gamma}(p, \mathcal{H}).$$

Because $p \in CR(\psi)$, by applying Claim 1 for ψ, we have $\mu(B \setminus \psi_{\mathbb{R}}(p)) = 0$. Hence $\mu(\psi_{\mathbb{R}}(p)) > 0$, which implies that p is a periodic point of ψ. Let $T > 1$ be such that $\psi_T(p) = p$. Since $x \in \Gamma^{\psi}_{2\gamma}(p, \mathcal{H})$, there exists $h \in \mathcal{H}$ satisfying $d(\psi_t(p), \psi_{h(t)}(x)) < 2\gamma$ for all $t \in \mathbb{R}$. Because Y is compact, the sequence $\{\psi_{h(nT)}(x)\}_{n\in\mathbb{Z}}$ has a limit point in $CR(\psi)$. We consider the following two cases.

Case 1 There exists a limit point x_0 of the sequence $\{\psi_{h(nT)}(x)\}_{n\in\mathbb{Z}}$ such that $x_0 \notin \psi_{\mathbb{R}}(p)$. Assume that there is a subsequence $\{n_k\}_{k\in\mathbb{N}}$ such that $\psi_{h(n_kT)}(x)$ converges to x_0 as $k \to \infty$. Note that $x_0 \in CR(\psi)$. For each $k \in \mathbb{N}$, define a homeomorphism $g_k : \mathbb{R} \to \mathbb{R}$ by

$$g_k(t) = h(n_kT + t) - h(n_kT).$$

Then, for all $t \in \mathbb{R}$,

$$d(\psi_{g_k(t)}(\psi_{h(n_kT)}(x)), \psi_t(p)) = d(\psi_{h(n_kT+t)}(x), \psi_{n_kT+t}(p)) \le 2\gamma.$$

Hence $\psi_{h(n_kT)}(x) \in \Gamma^{\psi}_{2\gamma}(p, \mathcal{H})$ for every $k \in \mathbb{N}$. Consequently, we get $x_0 \in \overline{\Gamma^{\psi}_{2\gamma}(p, \mathcal{H})} \subset \Gamma^{\psi}_{\delta}(p, \mathcal{H})$, and therefore $p \in \Gamma^{\psi}_{\delta}(x_0, \mathcal{H})$. Let $\mathcal{L}_p$ denote the invariant normalized Lebesgue measure supported on the periodic orbit $\psi_{\mathbb{R}}(p)$. Then

$$i(\mathcal{L}_p)(\Gamma^{\phi}_{\delta}(x_0) \setminus \phi_{\mathbb{R}}(x_0)) \ge \mathcal{L}_p(\Gamma^{\psi}_{\delta}(x_0, \mathcal{H}) \setminus \psi_{\mathbb{R}}(x_0)) > 0,$$

which contradicts Claim 1.

Case 2 Every limit point of the sequence $\{\psi_{h(nT)}(x)\}_{n\in\mathbb{Z}}$ belongs to the orbit $\psi_{\mathbb{R}}(p)$. Suppose there exists $s_1 < 0 < s_2$ such that $\psi_{s_1}(p)$ and $\psi_{s_2}(p)$ are limit points of $\{\psi_{h(nT)}(x)\}_{n>0}$ and $\{\psi_{h(nT)}(x)\}_{n<0}$, respectively. For any $\tau > 0$, choose integers $n_k > 0$ and $n_l < 0$ such that

$$h(n_kT) > 1, \quad -h(n_lT) > 1, \; d(\psi_{h(n_kT)}(x), \psi_{s_1}(p)) < \tau$$

and

$$d(\psi_{h(n_lT)}(x), \psi_{s_2}(p)) < \tau.$$

Then we can construct a $(\tau, 1)$-chain from x to itself as follows:

$$\{(x, h(n_k T)), (\psi_{s_1}(p), -s_1 + s_2 + T), (\psi_{h(n_l T)}(x), -h(n_l T))\}.$$

Since τ is arbitrary, it follows that $x \in CR(\psi)$. The contradiction shows that $\mu(\Gamma_\gamma^\psi(x) \setminus \psi_{\mathbb{R}}(x)) = 0$ for all $x \in Y$ and $\mu \in \mathcal{M}(Y, \psi)$. Because ψ has no fixed points, we may choose $0 < \varepsilon < \delta$ such that $\Gamma_\varepsilon^\psi(x) \subset \Gamma_\gamma^\psi(x, \mathcal{H})$ for all $x \in Y$. Therefore, ψ is invariant measure geometric expansive on Y with an invariant measure geometric expansive constant ε.

Claim 3: If $\phi_{\mathbb{R}}(x) \subset \Gamma_\delta^\phi(p)$ for $x \in X$ and $p \in Fix(\phi)$, then $x = p$.

Let $y \in \omega(x, \phi) \cup \alpha(x, \phi) \subset \Gamma_\delta^\phi(p) = \{y \in X | d(p, y) \le \delta\}$. Then $p \in \Gamma_\delta^\phi(y)$. Let δ_p denotes the Dirac measure centered at p. Since $\delta_p(\Gamma_\delta^\phi(y) \setminus \phi_{\mathbb{R}}(y)) = 0$, it follows that $y = p$. Hence, $\omega(x, \phi) \cup \alpha(x, \phi) = \{p\}$. For any $\tau > 0$, choose constants $T, S > 1$ such that

$$d(\phi_T(x), p) < \frac{\tau}{2} \quad \text{and} \quad d(\phi_{-S}(x), p) < \frac{\tau}{2}.$$

Then $\{(x, T), (\phi_{-S}(x), S)\}$ is a $(\tau, 1)$-chain of ϕ from x to itself. Since τ is arbitrary, x is a chain recurrent point of ϕ. Clearly, $p \in \Gamma_\delta^\phi(x)$, and so $\delta_p(\Gamma_\delta^\phi(x) \setminus \phi_{\mathbb{R}}(x)) = 0$. Therefore, $x = p$.

Claim 4: We show that $\mu(\Gamma_\varepsilon^\phi(x) \setminus \phi_{\mathbb{R}}(x)) = 0$ for any $x \in Y$ and $\mu \in \mathcal{M}(X, \phi)$.

Since ϕ is invariant measure geometric expansive on $CR(\phi)$, we see that $Fix(\phi)$ is finite, and denote it by $Fix(\phi) = \{p_i | i = 1, \ldots, n\}$. By Claim 1, for any $\mu \in \mathcal{M}(X, \phi)$, we have

$$\mu(U_\delta(Fix(\phi)) \setminus Fix(\phi)) \le \mu\left(\bigcup_{i=1}^{n} \Gamma_\delta^\phi(p_i) \setminus \{p_i\}\right) = 0.$$

Hence, μ contains a finite combination of Dirac measures δ_{p_i}, that is, $\sum_{i=1}^n \lambda_i \delta_{p_i}$, where $\lambda_i \ge 0$ and $0 \le \sum_{i=1}^n \lambda_i \le 1$. Moreover, we claim that $\mu(X \setminus (Y \cup Fix(\phi))) = 0$. Indeed, assume, to the contrary, that $\mu(X \setminus (Y \cup Fix(\phi))) > 0$. Then there exist a compact set $C \subset X \setminus (Y \cup Fix(\phi))$ with $\mu(C) > 0$ and a point $x \in C$ such that $\mu(B(x, r) \cap C) > 0$ for any $r > 0$. Since $x \in Y^c$, there exist $r > 0$ and $t \in \mathbb{R}$ such that $\phi_t(B(x, r) \cap C) \subset B(p, \delta)$ for some $p \in Fix(\phi)$. Because μ is invariant, we have

$$\mu(\Gamma_\delta^\phi(p) \setminus \{p\}) > \mu(B(x, r) \cap C) > 0,$$

which contradicts Claim 1. Consequently, $\mu(X \setminus (Y \cup Fix(\phi))) = 0$. Thus, every $\mu \in \mathcal{M}(X, \phi)$ can be written as

$$\mu(A) = \left(1 - \sum_{i=1}^{n} \lambda_i\right) i(\nu)(A) + \sum_{i=1}^{n} \lambda_i \delta_{p_i}(A)$$

for any Borel set A of X, where ν is an invariant measure on Y defined by

$$\nu(A) = \frac{\mu(A)}{1 - \sum_{i=1}^{n} \lambda_i}$$

for any Borel set A of Y. Now let $x \in Y$, and let $B \subset X$ be a Borel set such that $B \subset \Gamma_\varepsilon^\phi(x) \setminus \phi_\mathbb{R}(x)$. By Claim 2, we have $i(\nu)(B) = \nu(B \cap Y) = 0$. Since $B \cap Fix(\phi) = \emptyset$, we get $\delta_{p_i}(B) = 0$ for all $i = 1, \ldots, n$. Therefore,

$$\mu(B) = \left(1 - \sum_{i=1}^{n} \lambda_i\right) i(\nu)(B) + \sum_{i=1}^{n} \lambda_i \delta_{p_i}(B) = 0.$$

Hence, we get $\mu(\Gamma_\varepsilon^\phi(x) \setminus \phi_\mathbb{R}(x)) = 0$ for all $x \in Y$ and $\mu \in \mathcal{M}(X, \phi)$.

Claim 5: We show $\mu(\Gamma_\varepsilon^\phi(x) \setminus \phi_\mathbb{R}(x)) = 0$ for any $x \in Y^c$ and $\mu \in \mathcal{M}(X, \phi)$.

Assume, on the contrary, that there exist $x \in Y^c$ and $\mu \in \mathcal{M}(X, \phi)$ such that $\mu(B \setminus \phi_\mathbb{R}(x)) > 0$ for some Borel set $B \subset \Gamma_\varepsilon^\phi(x)$. By Claim 1, we have $x \notin CR(\phi)$. We consider the following two cases.

Case 1 $x \in \Gamma_\delta^\phi(p)$ for some $p \in Fix(\phi)$. In this case, $\Gamma_\varepsilon^\phi(x) \subset \Gamma_{2\delta}^\phi(p)$. Hence,

$$\mu(B \setminus (\phi_\mathbb{R}(x) \cup \phi_\mathbb{R}(p))) \leq \mu(\Gamma_{2\delta}^\phi(p) \setminus \phi_\mathbb{R}(p)) = 0.$$

This implies that $p \in B \subset \Gamma_\varepsilon^\phi(x)$, and therefore $d(\phi_t(x), p) \leq \varepsilon$ for all $t \in \mathbb{R}$. By Claim 3, it follows that $x = p \in CR(\phi)$, which is a contradiction.

Case 2 $x \notin \Gamma_\delta^\phi(p)$ for any $p \in Fix(\phi)$. Then $B \cap Fix(\phi) = \emptyset$. By Proposition 1.1, we may assume that B is closed. Hence, there exists $y \in B$ such that $\mu(B(y, \gamma) \cap B) > 0$ for all $\gamma > 0$. Since $\phi_t(x) \in B(p, \delta)$ for some $t \in \mathbb{R}$ and $p \in Fix(\phi)$, we can take $s \in \mathbb{R}$ and $\gamma > 0$ such that $\phi_s(B(y, \gamma) \cap B) \subset B(p, 2\delta) \subset \Gamma_{2\delta}^\phi(p)$. By Claim 1, we have

$$\mu(B(y, \gamma) \cap B) = \mu(\phi_s(B(y, \gamma) \cap B)) \leq \mu(\Gamma_{2\delta}^\phi(p) \setminus \phi_\mathbb{R}(p)) = 0.$$

The contradiction completes the proof.

□

In the following example, we construct a flow ϕ which is invariant measure geometric expansive on its nonwandering set $\Omega(\phi)$, but not invariant measure geometric expansive on X.

Example 9.3 For each $n \in \mathbb{N}$, let S_n be the circle in the plane $\mathbb{R}^2$ centered at $(0, \frac{1}{n})$ with radius $\frac{1}{n}$, and define $X = \bigcup_{n \in \mathbb{N}} S_n$. Consider a flow ϕ on the space X satisfying:

(1) $\mathbf{0} = (0, 0)$ is a unique fixed point of ϕ;
(2) for any $p = (x, y) \in S_n \setminus \{\mathbf{0}\}$, $\omega(p, \phi) = \alpha(p, \phi) = \{\mathbf{0}\}$ and $\overline{\phi_\mathbb{R}(p)} = S_n$.

Clearly, ϕ is invariant measure geometric expansive on $\Omega(\phi) = \{\mathbf{0}\}$. However, ϕ is not invariant measure geometric expansive on X. Indeed, for each $n \in \mathbb{N}$, let $x_n = (0, \frac{1}{n})$. For any $\delta > 0$, choose $n \in \mathbb{N}$ such that $\mathbf{0} \in \Gamma_\delta^\phi(x_n)$. Then

$$\delta_0(\Gamma_\delta^\phi(x_n) \setminus \phi_{\mathbb{R}}(x_n)) > 0,$$

where δ_0 denotes the Dirac measure centered at the origin $\mathbf{0}$, which is a ϕ-invariant probability measure on X. Hence, ϕ is not invariant measure geometric expansive on X.

9.3 Measurable Spectral Decomposition for Flows

In this section, we present a measurable version of Smale's spectral decomposition theorem for flows. More precisely, we prove that if a flow ϕ is invariant measure geometric expansive and invariant measure shadowing on its chain recurrent set $CR(\phi)$, then the flow ϕ has the spectral decomposition.

Theorem 9.4 *Let ϕ be a flow on a compact metric space X. If ϕ is invariant measure geometric expansive and invariant measure shadowing on its chain recurrent set $CR(\phi)$, then ϕ has the spectral decomposition, i.e., the nonwandering set $\Omega(\phi)$ is decomposed by a disjoint union of finitely many invariant and closed subsets*

$$\Omega(\phi) = \Omega_1 \cup \cdots \cup \Omega_m$$

such that ϕ is topologically transitive on each Ω_i for $1 \le i \le m$.

Before proving the theorem, we give an example to show that a measure expansive flow with shadowing does not admit the spectral decomposition. Note that the homeomorphism f in Example 9.2 is measure expansive, and shadowing. However, since f has infinitely many chain components which are open in its nonwandering set $\Omega(f)$, we see that it does not have the spectral decomposition.

Example 9.4 Let ϕ be the suspension flow under the constant height function $\tau = 1$ of the homeomorphism f on $X = \Sigma_2 \cup F$, where $F = \{q_{nk} : n \in \mathbb{N} \text{ and } 0 \le k < n\}$ as described in Example 9.2. Since f is 2-expansive and shadowing, it follows that the flow ϕ is measure expansive and shadowing.

We observe that ϕ does not admit the spectral decomposition. Suppose, on the contrary, that ϕ has the spectral decomposition, that is, the nonwandering set $\Omega(\phi)$ can be written as a disjoint union of finitely many invariant and closed subsets

$$\Omega(\phi) = \Omega_1 \cup \cdots \cup \Omega_m,$$

each of which is topologically transitive under ϕ. Then there is a component Ω_i $(1 \le i \le m)$ that contains infinitely many periodic orbits $\phi_{\mathbb{R}}((q_{n_j0}, 0))$ for $j \in \mathbb{N}$. Moreover, each orbit $\phi_{\mathbb{R}}((q_{n_j0}, 0))$ is open in X^f. Since ϕ is topologically transitive on Ω_i, for any distinct $j, k \in \mathbb{N}$, there exists $T \ge 1$ such that

$$\phi_T(\phi_{\mathbb{R}}((q_{n_j0}, 0))) \cap \phi_{\mathbb{R}}((q_{n_k0}, 0)) \neq \emptyset.$$

This leads to a contradiction, and hence ϕ does not admit the spectral decomposition.

To prove Theorem 9.4, we need a proposition and several lemmas. We first observe that if a flow ϕ on X is expansive then every fixed point of ϕ is isolated in X (see Lemma 6.1). However, this is not true for invariant measure geometric expansive flows as we can see in Example 9.1.

In the following proposition, we see that every fixed point of an invariant measure geometric expansive flow ϕ is isolated in $CR(\phi)$ if it is shadowing on $CR(\phi)$.

Proposition 9.2 *If a flow ϕ is invariant measure geometric expansive and shadowing on $CR(\phi)$, then every fixed point of ϕ is isolated in $CR(\phi)$.*

Proof Suppose that ϕ is invariant measure geometric expansive and shadowing on $CR(\phi)$. Then, by Theorem 9.3, ϕ is invariant measure geometric expansive on X . It is also clear that ϕ admits only finitely many fixed points.

Let ε be an invariant measure geometric expansive constant of ϕ. We will show that for any $p \in Fix(\phi)$,

$$B\big(p, \tfrac{\varepsilon}{2}\big) \cap CR(\phi) = \{p\}.$$

Suppose, on the contrary, there exists $x \neq p$ such that $x \in B(p, \varepsilon/2) \cap CR(\phi)$. choose a constant $\delta > 0$ corresponding to $\min\{d(x, p)/2, \varepsilon/4\}$ by the shadowing of ϕ on $CR(\phi)$. Since $x \in CR(\phi)$, there is a $(\delta, 1)$-pseudo orbit $\{(x_i, t_i)\}_{i=0}^{n}$ from x to itself which is contained in $CR(\phi)$. We extend it to a periodic $(\delta, 1)$-pseudo orbit $\xi = \{(x_i, t_i)\}_{i\in\mathbb{Z}}$ through x by letting $x_i = x_{i+kn}$ and $t_i = t_{i+kn}$ for all $k \in \mathbb{Z}$ and $0 \le i \le n-1$. Then there are $y \in X$ and $h \in \mathcal{H}$ such that

$$d\big(\phi_{h(t)}(y), \phi_{t-T_i}(x_i)\big) \le \tfrac{\varepsilon}{4}, \quad \forall t \in [T_i, T_{i+1}), \ i \in \mathbb{Z},$$

where T_i is the ith sum associated to ξ. Since

$$d(p, y) \ge d(p, x) - d(x, y) \ge \tfrac{d(p,x)}{2} > 0,$$

we have $p \neq y$. Moreover, we get

$$d(y, p) \le d(y, x) + d(x, p) < \varepsilon.$$

Case 1 Suppose that y is a periodic point of ϕ, and denote by $\mathcal{L}_y$ the invariant normalized Lebesgue measure supported on the periodic orbit of y. Then we have $\mathcal{L}_y(\Gamma_\varepsilon^\phi(p) \setminus \phi_{\mathbb{R}}(p)) > 0$, which is a contradiction.

Case 2 Suppose y is not periodic for ϕ, and let $T = \sum_{i=0}^{n-1} t_i$. For each $k \in \mathbb{Z}$, we have

$$d(y, \phi_{h(kT)}(y)) \le d(y, x_0) + d(x_0, \phi_{h(kT)}(y)) < \frac{3\varepsilon}{4}.$$

Hence, $\phi_{h(kT)}(y) \in B(y, 3\varepsilon/4)$ for all $k \in \mathbb{Z}$. Since y is not periodic for ϕ, we can define a continuous map $f : \mathbb{R} \to X$ by $f(t) = \phi_{h(t)}(y)$ for all $t \in \mathbb{R}$. For each $i \in \mathbb{N}$, let

$$A_i = \{\phi_{h(t)}(y) | t \in [-iT, iT]\}.$$

We now define a sequence of Borel measures $\{\mu_i\}_{i\in\mathbb{N}}$ on X by

$$\mu_i(A) = \frac{1}{2iT} m(f^{-1}(A \cap A_i)),$$

for any Borel set A in X, where m denotes the Lebesgue measure on $\mathbb{R}$. Because X is compact, there exists $\alpha > 0$ such that

$$\phi_{h(t+kT)}(y) \in B(y, \varepsilon), \quad \forall t \in [-\alpha, \alpha] \text{ and } \forall k \in \mathbb{Z}.$$

Consequently,

$$\mu_i(B(y, \varepsilon)) = \frac{1}{2iT} m(f^{-1}(B(y, \varepsilon) \cap A_i)) \geq \frac{1}{2iT} \cdot 2i\alpha = \frac{\alpha}{T},$$

for all $i \in \mathbb{N}$. Let μ be the weak* limit of a subsequence $\{\mu_{i_j}\}_{j\in\mathbb{N}}$ in $\mathcal{M}(X)$. Then we have

$$\mu(\overline{B(y, \varepsilon)}) \geq \limsup_{j\to\infty} \mu_{i_j}(\overline{B(y, \varepsilon)}) \geq \frac{\alpha}{T} > 0.$$

Since $\overline{B(y, \varepsilon)} \subset \Gamma_\varepsilon^\phi(p)$, we get $\mu(\overline{B(y, \varepsilon)} \setminus \phi_{\mathbb{R}}(p)) = 0$, and so $y = p$. This contradiction completes the proof.

□

Consequently, we have the following corollary if we apply Theorem 6.4 and Proposition 9.2.

Corollary 9.1 *Let ϕ be a flow which is invariant measure geometric expansive on its chain recurrent set $CR(\phi)$. Then ϕ is shadowing on $CR(\phi)$ if and only if it is strongly shadowing on $CR(\phi)$.*

For any $\varepsilon, \delta > 0$ and $x \in X$, we define

$$\Gamma_\delta^\phi(x, \varepsilon) = \{y \in X | d(\phi_t(x), \phi_{h(t)}(y)) \leq \delta \text{ for some } h \in \mathcal{H}(\varepsilon) \text{and all } t \in \mathbb{R}\}.$$

Lemma 9.3 *Let ϕ be a flow on a compact metric space X. Then for any $\varepsilon, \delta > 0$, we have:*

(1) $\Gamma_\delta^\phi(x, \varepsilon)$ is closed for $x \in X$;
(2) for any $T > 0$ and $x \in X$, there is $\{l_i\}_{i=1}^n$ with $0 \leq l_i \leq T$ $(i = 1, \ldots, n)$ such that

$$\bigcup_{0\leq l\leq T} \Gamma_\delta^\phi(\phi_l(x), \varepsilon) \subset \bigcup_{i=1}^n \Gamma_{2\delta}^\phi(\phi_{l_i}(x_i), \varepsilon).$$

Proof (1) Let $\varepsilon, \delta > 0$, and $\{x_n\}_{n\in\mathbb{N}}$ be a sequence in $\Gamma_\delta^\phi(x,\varepsilon)$ which converges to y for some $y \in X$. For each $n \in \mathbb{N}$, there is $h_n \in \mathcal{H}(\varepsilon)$ such that $d(\phi_t(x), \phi_{h_n(t)}(x_n)) \leq \delta$ for all $t \in \mathbb{R}$. It is clear that $\{h_n\}_{n\in\mathbb{N}}$ is pointwise bounded and equicontinuous. By the Arzela–Ascoli theorem, there is a homeomorphism h of $\mathbb{R}$ such that for any $N > 0$ there is a subsequence $\{h_{n_k}\}_{k\in\mathbb{N}}$ of $\{h_n\}_{n\in\mathbb{N}}$ such that $h_{n_k}|_{[-N,N]}$ uniformly converges to $h|_{[-N,N]}$ as k tends to infinity. Since $h_{n_k} \in \mathcal{H}(\varepsilon)$, we can check that $h \in \mathcal{H}(\varepsilon)$.

For any $\gamma > 0$, we claim that $d(\phi_t(x), \phi_{h(t)}(y)) \leq \delta + \gamma$ for all $t \in \mathbb{R}$. Indeed, for each $t \in \mathbb{R}$, there are $N > 0$ such that $t \in [-N, N]$, and $h_{n_k}|_{[-N,N]}$ uniformly converges to $h|_{[-N,N]}$ as k tends to infinity. Since

$$d(\phi_{h_{n_k}(t)}(y), \phi_{h_{n_k}(t)}(x_{n_k})) \to 0 \text{ as } k \to \infty,$$

we take $l \in \mathbb{N}$ such that

$$d(\phi_{h_{n_l}(t)}(y), \phi_{h_{n_l}(t)}(x_{n_l})) < \gamma/2 \text{ and } d(\phi_{h(t)}(y), \phi_{h_{n_l}(t)}(y)) < \gamma/2.$$

Then we have

$$\begin{aligned} d(\phi_t(x), \phi_{h(t)}(y)) &\leq d(\phi_t(x), \phi_{h_{n_l}(t)}(x_{n_l})) + d(\phi_{h_{n_l}(t)}(x_{n_l}), \phi_{h_{n_l}(t)}(y)) \\ &\quad + d(\phi_{h_{n_l}(t)}(y), \phi_{h(t)}(y)) \\ &< \delta + \frac{\gamma}{2} + \frac{\gamma}{2} = \delta + \gamma, \end{aligned}$$

and so $d(\phi_t(x), \phi_{h(t)}(y)) < \delta + \gamma$ for all $t \in \mathbb{R}$.

Since γ is arbitrary, we get $d(\phi_t(x), \phi_{h(t)}(y)) \leq \delta$ for all $t \in \mathbb{R}$. Hence we derive that $y \in \Gamma_e^\phi(x,\varepsilon)$, and so $\Gamma_e^\phi(x,\varepsilon)$ is closed.

(2) Let $T > 0$ and $x \in X$. For given $\delta > 0$, by the continuity of ϕ, there is $\alpha > 0$ such that $d(\phi_t(x), \phi_t(\phi_s(x))) \leq \delta$ for all $s \in (-\alpha, \alpha)$ and $t \in \mathbb{R}$. For each $l \in [0, T]$, we see that $\phi_{(-\alpha,\alpha)}(\phi_l(x)) \subset \Gamma_\delta^\phi(\phi_l(x), \varepsilon)$. Since $\phi_{[0,T]}(x)$ is compact and

$$\phi_{[0,T]}(x) \subset \bigcup_{0\leq l\leq T} \phi_{(-\alpha,\alpha)}(\phi_l(x)),$$

there are $\{l_i\}_{i=1}^n$ with $0 \leq i \leq T$ such that

$$\phi_{[0,T]}(x) \subset \bigcup_{i=1}^{n} \phi_{(-\alpha,\alpha)}(\phi_{l_i}(x)).$$

Let $y \in \bigcup_{0\leq l\leq T} \Gamma_\delta^\phi(\phi_l(x), \varepsilon)$. We suppose that $y \in \Gamma_\delta^\phi(\phi_l(x), \varepsilon)$ for some $l \in [0, T]$. For such $l \in [0, T]$, we have $\phi_l(x) \in \phi_{(-\alpha,\alpha)}(\phi_{l_i}(x))$ for some $i \in$

$\{1, \ldots, n\}$, and so $\phi_l(x)$ can be written as $\phi_l(x) = \phi_{l_i+s}(x)$ for some $s \in (-\alpha, \alpha)$. By the choice of α, we have $\phi_l(x) \in \Gamma_\delta^\phi(\phi_{l_i}(x), \varepsilon)$. Since $y \in \Gamma_\delta^\phi(\phi_l(x), \varepsilon)$, there is $h \in \mathcal{H}(\varepsilon)$ such that $d(\phi_{h(t)}(y), \phi_{t+l}(x)) \leq \varepsilon$ for all $t \in \mathbb{R}$. Since

$$\begin{aligned} d(\phi_{h(t)}(y), \phi_t(\phi_{l_i}(x))) &\leq d(\phi_{h(t)}(y), \phi_t(\phi_l(x))) + d(\phi_t(\phi_l(x)), \phi_t(\phi_{l_i}(x))) \\ &\leq \delta + \delta = 2\delta, \ \forall t \in \mathbb{R}, \end{aligned}$$

we have $y \in \Gamma_{2e}^\phi(\phi_{l_i}(x), \varepsilon)$. Hence we get

$$\bigcup_{0 \leq l \leq T} \Gamma_e^\phi(\phi_l(x), \varepsilon) \subset \bigcup_{i=1}^{n} \Gamma_{2e}^\phi(\phi_{l_i}(x_i), \varepsilon).$$

This completes the proof.

□

Lemma 9.4 *Let ϕ be a flow on a compact metric space X. If ϕ is invariant measure geometric expansive and shadowing on $CR(\phi)$, then $CR(\phi) = \overline{Per(\phi)}$.*

Proof Suppose that ϕ is invariant measure geometric expansive and shadowing on $CR(\phi)$. By Theorems 8.2, 9.3, and Corollary 9.1, it follows that ϕ is invariant measure geometric expansive on X and strongly shadowing on $CR(\phi)$. Let $e > 0$ be an invariant measure geometric expansive constant of ϕ. For any $0 < \varepsilon < \min\{e/4, 1\}$, since $\phi|_{CR(\phi)}$ is strongly shadowing, we take a constant $\delta > 0$ such that every $(\delta, 1)$-pseudo orbit of ϕ in $CR(\phi)$ is strongly ε-shadowed by some point in $CR(\phi)$.

Fix $x \in CR(\phi)$, and consider a $(\delta, 1)$-pseudo orbit $\{(x_i, t_i)\}_{i=0}^m$ of ϕ in $CR(\phi)$ from x to itself. We extend this to a periodic $(\delta, 1)$-pseudo orbit $\xi = \{(x_i, t_i)\}_{i \in \mathbb{Z}}$ through x by setting $x_i = x_{i+km}$ and $t_i = t_{i+km}$ for all $k \in \mathbb{Z}$ and $0 \leq i \leq m-1$. Then, by the strong shadowing of $\phi|_{CR(\phi)}$, there exist $p \in X$ and $h \in \mathcal{H}(\varepsilon)$ such that

$$d(\phi_{h(t)}(p), \phi_{t-T_i}(x_i)) < \varepsilon, \ \forall t \in [T_i, T_{i+1}),$$

where T_i is the ith sum associated to ξ. Let $T = \sum_{i=0}^{m-1} t_i$. Then for any $l \in [0, T]$ and $k \in \mathbb{Z}$, we have

$$\begin{aligned} d(\phi_{h(t+l)}(p), \phi_{h(t+kT+l)}(p)) &\leq d(\phi_{h(t+l)}(p), \phi_{t-T_i}(x_i)) \\ &\quad + d(\phi_{t-T_i}(x_i), \phi_{h(t+kT+l)}(p)) < 2\varepsilon, \ \ \forall t \in \mathbb{R}. \end{aligned}$$

Define homeomorphisms $f, g : \mathbb{R} \to \mathbb{R}$ by

$$f(t) = h(t+l) - h(l) \ \text{ and } \ g(t) = h(t+l+kT) - h(l+kT), \quad \forall t \in \mathbb{R}.$$

Then, for any $t \in \mathbb{R}$ and $s = f^{-1}(t)$, we have

$$d\big(\phi_t(\phi_{h(l)}(p)), \phi_{g\circ f^{-1}(t)}(\phi_{h(l+kT)}(p))\big) = d\big(\phi_{h(s+l)}(p), \phi_{h(s+kT+l)}(p)\big) < 2\varepsilon.$$

Let $\tau = 2\varepsilon/(1-\varepsilon)$. Then we get $g \circ f^{-1} \in \mathcal{H}(\tau)$, and so we have

$$\phi_{h(l+kT)}(p) \in \Gamma_{2\varepsilon}^{\phi}(\phi_{h(l)}(p), \tau)$$

for $l \in [0, T]$ and $k \in \mathbb{Z}$. By Lemma 9.3, there is $\{l_i\}_{i=1}^{n}$ such that

$$\overline{\phi_{\mathbb{R}}(p)} \subset \overline{\bigcup_{l\in[0,h(T)]} \Gamma_{2\varepsilon}^{\phi}(\phi_l(p), \tau)} \subset \bigcup_{i=1}^{n} \Gamma_{4\varepsilon}^{\phi}(\phi_{l_i}(p), \tau).$$

Let μ be an invariant Borel measure supported on $\overline{\phi_{\mathbb{R}}(p)}$. Then there exists $1 \leq i \leq n$ such that

$$\mu\big(\Gamma_{4\varepsilon}^{\phi}(\phi_{l_i}(p), \tau)\big) > 0.$$

Since ϕ is invariant measure geometric expansive, we obtain

$$\mu\big(\Gamma_{4\varepsilon}^{\phi}(\phi_{l_i}(p), \tau) \setminus \phi_{\mathbb{R}}(p)\big) = 0.$$

Hence $\mu(\phi_{\mathbb{R}}(p)) > 0$, and therefore p is periodic. Because $d(x, p) < \varepsilon$ and $\varepsilon > 0$ is arbitrary, we conclude that $x \in \overline{Per(\phi)}$. This completes the proof.

□

Lemma 9.5 *Let ϕ be an invariant measure geometric expansive flow on a compact metric space X. Then there exists a constant $e > 0$ such that if*

$$d(\phi_{g(t)}(y), \phi_t(p)) < e \textit{ for all } t > 0 \textit{ [resp.} t < 0],$$

for some $p \in Per(\phi)$, $y \in X$ and $g \in \mathcal{H}(e)$, then

$$d(\phi_{g(t)}(y), \phi_{\mathbb{R}}(p)) \to 0 \textit{ as } t \to \infty \textit{ [resp.} t \to -\infty].$$

Proof Let $e > 0$ be an invariant measure geometric expansive of ϕ, and let $p \in Per(\phi)$ and $y \in X$ satisfy

$$d(\phi_{g(t)}(y), \phi_t(p)) < e \text{ for all } t > 0,$$

for some $g \in \mathcal{H}(e)$. Suppose, to the contrary, that there exist a sequence $\{t_n\}_{n\in\mathbb{N}}$ in $\mathbb{R}$ such that $t_n \to \infty$ as $n \to \infty$, and a constant $r > 0$ such that

$$d(\phi_{g(t_n)}(y), \phi_{\mathbb{R}}(p)) > r \text{ for all} n \in \mathbb{N}.$$

Since X is compact, we may assume that $\phi_{t_n}(p) \to p_0$ and $\phi_{g(t_n)}(y) \to y_0$ for some $p_0 \in \phi_{\mathbb{R}}(p)$ and $y_0 \in X$. Then $d(y_0, \phi_{\mathbb{R}}(p)) \geq r$, hence $y_0 \notin \phi_{\mathbb{R}}(p)$. For each $n \in \mathbb{N}$, define a homeomorphism $h_n \in \mathcal{H}(e)$ by

$$h_n(t) = g(t_n + t) - g(t_n), \quad \forall t \in \mathbb{R}.$$

The family $\{h_n\}_{n\in\mathbb{N}}$ is pointwise bounded and equicontinuous. By the Arzela–Ascoli theorem, there exists a subsequence $\{h_{n_k}\}_{k\in\mathbb{N}}$ of $\{h_n\}_{n\in\mathbb{N}}$ which converges compactly to a homeomorphism h of $\mathbb{R}$. Fix $t \in \mathbb{R}$. Choose $N > 0$ and $k_0 \in \mathbb{N}$ such that $t \in [-N, N]$. Then we have

$$\begin{aligned} d(\phi_t(p_0), \phi_{h(t)}(y_0)) &= \lim_{k\to\infty} d(\phi_t(\phi_{t_{n_k}}(p)), \phi_{h_{n_k}(t)}(\phi_{g(t_{n_k})}(y))) \\ &= \lim_{k\to\infty} d(\phi_{t+t_{n_k}}(p), \phi_{g(t+t_{n_k})}(y)) \le e. \end{aligned}$$

Since t is arbitrary, it follows that

$$d(\phi_{h^{-1}(t)}(p_0), \phi_t(y_0)) \le e \text{ for all } t \in \mathbb{R}.$$

Hence $p_0 \in \Gamma_e^{\phi}(y_0)$. Let $\mathcal{L}_{p_0}$ be the invariant normalized Lebesgue measure supported on the periodic orbit $\phi_{\mathbb{R}}(p_0)$. Since $y_0 \notin \phi_{\mathbb{R}}(p_0)$, we have $\mathcal{L}_{p_0}(\Gamma_e^{\phi}(y_0) \setminus \phi_{\mathbb{R}}(y_0)) > 0$, which is a contradiction.

A similar argument shows that if $d(\phi_{g(t)}(y), \phi_t(p)) < e$ for all $t < 0$ for some $p \in Per(\phi)$, $y \in X$ and $g \in \mathcal{H}(e)$, then $d(\phi_{g(t)}(y), \phi_{\mathbb{R}}(p)) \to 0$ *as* $t \to -\infty$. This completes the proof.

□

We now present a measurable version of Smale's spectral decomposition theorem. More precisely, we prove that if a flow ϕ on a compact metric space is invariant measure geometric expansive on its chain recurrent set $CR(\phi)$ and is invariant measure shadowing on $CR(\phi)$ then ϕ has the spectral decomposition.

Proof of Theorem 9.4 Suppose that ϕ is invariant measure geometric expansive and invariant measure shadowing on its chain recurrent set $CR(\phi)$. Then, by Theorems 6.4, 8.2, and 9.3, it follows that ϕ is invariant measure geometric expansive on X, and that the restriction $\phi|_{\Omega(\phi)}$ is strongly shadowing. Moreover, by Lemma 9.4, we have $CR(\phi) = \overline{Per(\phi)}$. Consequently, the nonwandering set $\Omega(\phi)$ can be decomposed into the union of its chain components

$$\Omega(\phi) = \bigcup_{\lambda\in\Lambda} \Omega_\lambda,$$

where each Ω_λ is a chain component of ϕ, invariant under ϕ, and closed in $\Omega(\phi)$.

We first show that each Ω_λ is open in $\Omega(\phi)$. Let $e > 0$ be an invariant measure geometric expansive constant for ϕ. By the strong shadowing of $\phi|_{\Omega(\phi)}$, choose $\delta > 0$ such that every $(\delta, 1)$-pseudo orbit of ϕ in $CR(\phi)$ is strongly e-shadowed by some point in $\Omega(\phi)$. For each $\lambda \in \Lambda$, define a neighborhood U_λ of Ω_λ in $\Omega(\phi)$ by

$$U_\lambda = \{x \in \Omega(\phi) \mid d(x, \Omega_\lambda) < \delta/2\}.$$

We claim that if $p \in Per(\phi) \cap U_\lambda$, then $p \in \Omega_\lambda$. Indeed, take $y \in \Omega_\lambda$ such that $d(y, p) < \delta/2$. For any $0 < \varepsilon < \delta/2$, choose a periodic point p_ε satisfying $d(y, p_\varepsilon) < \varepsilon$. By Lemma 6.5, it follows that $p_\varepsilon \sim^\varepsilon y$.

To show that $p_\varepsilon \sim^\varepsilon p$, choose constants $T, S > 1$ such that $\phi_T(p) = p$ and $\phi_S(p_\varepsilon) = p_\varepsilon$. Consider the sequence $\{(x_i, t_i)\}_{i \in \mathbb{Z}}$ defined by

$$(x_i, t_i) = \begin{cases} (\phi_{iT}(p),\ T), & i \geq 0, \\ (\phi_{iS}(p_\varepsilon),\ S), & i < 0. \end{cases}$$

Since

$$d(p, p_\varepsilon) \leq d(p, y) + d(y, p_\varepsilon) \leq \delta,$$

it follows that $\xi = \{(x_i, t_i)\}_{i \in \mathbb{Z}}$ is a $(\delta, 1)$-pseudo orbit contained in $\Omega(\phi)$. Hence there exist $x \in \Omega(\phi)$ and $g \in \mathcal{H}(e)$ such that

$$d(\phi_{g(t)}(x), \phi_{t-T_i}(x_i)) < e \quad \text{for all } T_i \leq t < T_{i+1},\ i \in \mathbb{Z},$$

where T_i denotes the i-th sum associated to ξ. By Lemma 9.5, we have

$$d(\phi_{g(t)}(x), \phi_{\mathbb{R}}(p)) \to 0 \quad \text{as } t \to \infty,$$

and similarly,

$$d(\phi_{g(t)}(x), \phi_{\mathbb{R}}(p_\varepsilon)) \to 0 \quad \text{as } t \to -\infty.$$

Therefore, we can construct $(\varepsilon, 1)$-chains in Ω_λ from p_ε to p and from p to p_ε, which implies $p_\varepsilon \sim^\varepsilon p$. Since $\varepsilon > 0$ is arbitrary, we conclude that $p \sim y$, and hence $p \in \Omega_\lambda$.

Consequently,

$$U_\lambda = U_\lambda \cap \overline{Per(\phi)} \subset \overline{U_\lambda \cap Per(\phi)} \subset \Omega_\lambda.$$

This shows that Ω_λ is open in $\Omega(\phi)$. Since $\Omega(\phi)$ is compact, it follows that

$$\Omega(\phi) = \bigcup_{i=1}^{m} \Omega_i,$$

a finite disjoint union of chain components.

Next, we show that ϕ is topologically transitive on Ω_i for each $1 \leq i \leq m$. Let U, V be nonempty open subsets of Ω_i. Choose $r > 0$ such that $d_H(\Omega_k, \Omega_l) > r$ for all distinct $k, l \in \{1, \ldots, m\}$, and such that $B(x, r) \cap \Omega_i \subset U$ and $B(y, r) \cap \Omega_i \subset V$ for some $x \in U$ and $y \in V$.

By the strong shadowing of $\phi|_{\Omega(\phi)}$, choose $\gamma > 0$ such that every $(\gamma, 1)$-pseudo orbit of ϕ in $\Omega(\phi)$ is strongly r-shadowed by a point in $\Omega(\phi)$. Since $x, y \in \Omega_i$, there exists a $(\gamma, 1)$-pseudo orbit $\{(w_i, t_i)\}_{i=0}^{n}$ in Ω_i with $w_0 = x$ and $w_n = y$. We extend this to a $(\gamma, 1)$-pseudo orbit $\chi = \{(y_i, s_i)\}_{i\in\mathbb{Z}}$ in Ω_i by defining

$$(y_i, s_i) = \begin{cases} (\phi_i(x), 1), & i < 0, \\ (w_i, t_i), & 0 \le i \le n, \\ (\phi_{i-n}(y), 1), & i > n. \end{cases}$$

Then there exist $z \in \Omega(\phi)$ and $h \in \mathcal{H}(r)$ such that

$$d(\phi_{h(t)}(z), \phi_{t-T_i}(y_i)) < r \quad \text{for all } T_i \le t < T_{i+1},\ i \in \mathbb{Z},$$

where T_i denotes the i-th sum associated to χ. Since $z \in \Omega(\phi)$ and $d(z, \Omega_i) \le d(z, x) < r$, we have $z \in \Omega_i$. Moreover, as $d(x, z) < r$ and $d(\phi_{h(T_n)}(z), y) < r$, it follows that

$$\phi_{h(T_n)}(B(x, r)) \cap B(y, r) \cap \Omega_i \subset \phi_{h(T_n)}(U) \cap V \neq \emptyset.$$

Hence, ϕ is topologically transitive on Ω_i.

□

References

1. Akin, E.: The general topology of dynamical systems. In: Graduate Studies in Mathematics, vol. 1. American Mathematical Society, Providence, RI (1993)
2. Aoki, N.: On homeomorphisms with pseudo-orbit tracing property. Tokyo J. Math. **6**, 329–334 (1983)
3. Aoki, N., Hiraide, K.: Topological theory of dynamical systems. In: Recent Advances. North-Holland Mathematical Library, vol. 52. North-Holland Publishing Co., Amsterdam (1994)
4. Aponte, F., Villavicencio, H.: Shadowable points for flows. J. Dyn. Cont. Syst. **24**, 701–719 (2018)
5. Arbieto, A., Morales, C.A.: Some properties of positive entropy maps. Ergodic Theory Dyn. Syst. **34**, 765–776 (2014)
6. Artigue, A.: Kinematic expansive flows. Ergodic Theory Dyn. Syst. **36**, 390–421 (2016)
7. Artigue, A., Carrasco-Olivera, D.: A note on measure-expansive diffeomorphisms. J. Math. Anal. Appl. **428**, 712–716 (2015)
8. Artigue, A., Pacifico, M.J., Vieitez, J.: N-expansive homeomorphisms in surface. Commun. Contemp. Math. **18**, 1650040 (2016)
9. Bogachev, V.I.: Measure Theory, vol. II. Springer-Verlag, Berlin (2007)
10. Bowen, R.: Equilibrium states and ergodic theory of Axiom *A* diffeomorphisms. In: Lecture Notes in Mathematice, vol. 470. Springer-Verlag (1975)
11. Bowen, R., Walters, P.: Expansive one-parameter flows. J. Diff. Eq. **12**, 180–193 (1972)
12. Brin, M., Katok, A.: On Local Entropy, Geometric Dynamics. Lecture Notes in Mathematics, vol. 1007, pp. 30–38. Springer, Berlin (1983)
13. Carrasco-Olivera, D., Morales, C.A.: Expansive measures for flows. J. Differ. Equ. **256**, 2246–2260 (2014)
14. Carvalho, B., Cordeiro, W.: N-expansive homeomorphisms with the shadowing property. J. Differ. Equ. **261**, 3734–3755 (2016)
15. Cerminara, M., Lewowicz, J.: Some open problems concerning expansive systems. Rend. Istit. Mat. Univ. Trieste **42**, 129–141 (2010)
16. Cordeiro, W., Denker, M., Zhang, X.: On specification and measure expansiveness. Disc. Contin. Dyn. Syst. **37**, 1941–1957 (2017)
17. Dankner, A.: On Smale's axiom *A* dynamical systems. Ann. Math. **2**(107), 517–553 (1978)
18. Das, T., Lee, K., Richeson, D., Wiseman, J.: Spectral decomposition for topologically Anosov homeomorphisms on noncompact and non-metrizable spaces. Topol. Appl. **160**, 149–158 (2013)
19. Gutiérrez, C.: Smoothing continuous flows on two-manifolds and recurrences. Ergodic Theory Dyn. Syst. **6**, 17–44 (1986)

K. Lee et al., *Topological Dynamics from a Measure-Theoretic Viewpoint*, Springer Asia Pacific Mathematics Series 14, https://doi.org/10.1007/978-981-95-6390-6

20. He, L.F., Shan, G.Z.: The nonexistence of expansive flow on a compact 2-manifold. Chin. Ann. Math. Ser. B **12**, 213–218 (1991)
21. Hector, G., Hirsch, U.: Introduction to the geometry of foliations. Part B. In: Foliations of Codimension One, 2nd edn. Aspects of Mathematics, vol. E3. Friedr. Vieweg & Sohn, Braunschweig (1987)
22. Helmuth, V.F.: $\mathcal{F}$-expansivity for Borel measures. J. Differ. Equ. **261**, 5350–5370 (2016)
23. Katok, A., Hasselblatt, B.: Introduction to the modern theory of dynamical systems. With a supplementary chapter by Katok and Leonardo Mendoza. In: Encyclopedia of Mathematics and its Applications, vol. 54. Cambridge University Press, Cambridge (1995)
24. Kawaguchi, N.: Quantitative shadowable points. Dyn. Syst. **32**, 504–518 (2017)
25. Kawaguchi, N.: Properties of shadowable points: chaos and equicontinuity. Bull. Braz. Math. Soc. **48**, 599–622 (2017)
26. Komuro, M.: One-parameter flows with the pseudo-orbit tracing property. Monatshefte Math. **98**, 219–253 (1984)
27. Lee, K., Morales, C.A.: Topological stability and pseudo-orbit tracing property for expansive measures. J. Differ. Equ. **262**, 3467–3487 (2017)
28. Lee, K., Nguyen, N.: Spectral decomposition and Ω-stability of flows with expanding measures. J. Differ. Equ. **269**, 7574–7604 (2020)
29. Lee, K., Nguyen, N., Yang, Y.: Topological stability and spectral decomposition for homeomorphisms on noncompact spaces. Disc. Contin. Dyn. Syst. **38**, 2487–2503 (2018)
30. Luzzatto, S., Melbourne, I., Paccaut, F.: The Lorenz attractor is mixing. Commun. Math. Phys. **260**, 393–401 (2005)
31. Palis, J., de Melo, W.: Geometric Theory of Dynamical Systems. An Introduction. Springer-Verlag, New York-Berlin (1982)
32. Parthasarathy, K.R.: Probability measures on metric spaces. In: Probability and Mathematical Statistics. Academic Press Inc, New York-London (1967)
33. Pilyugin, S.Y.: The space of dynamical systems with the C^0-topology. In: Lecture Notes in Mathematics, vol. 1571. Springer-Verlag, Berlin (1994)
34. Pilyugin, S.Y., Plamenevskaya, O.B.: Shadowing is generic. Topol. Appl. **97**, 253–266 (1999)
35. Ruggiero, R.O.: Expansive dynamics and hyperbolic geometry. Bol. Soc. Brasil. Mat. **25**, 139–172 (1994)
36. Smale, S.: Differentiable dynamical systems. Bull. Am. Math. Soc. **73**, 747–817 (1967)
37. Thomas, R.F.: Stability properties of one-parameter flows. Proc. Lond. Math. Soc. **45**, 479–505 (1982)
38. Viana, M.: Dynamical systems: moving into the next century. In: Mathematics Unlimited 2002 and Beyond, pp. 1167–1178. Springer, Berlin (2001)
39. Walters, P.: On the pseudo-orbit tracing property and its relationship to stability. In: The Structure of Attractors in Dynamical Systems. Lecture Notes in Mathematics, vol. 668, pp. 231–244. Springer, Berlin (1978)

Index

K. Lee et al., *Topological Dynamics from a Measure-Theoretic Viewpoint*, Springer Asia Pacific Mathematics Series 14, https://doi.org/10.1007/978-981-95-6390-6

www.ingramcontent.com/pod-product-compliance
Ingram Content Group UK Ltd.
Pitfield, Milton Keynes, MK11 3LW, UK
UKHW021616260926
475829UK00011B/107
* 9 7 8 9 8 1 9 5 6 3 8 9 0 *